Armoured Vehicles of the Iraq War

Finlay Reynolds

MILITARY VEHICLES AND ARTILLERY SERIES, VOLUME 10

Front cover image: The Challenger 2 Main Battle Tank (MBT) was first used in combat in March 2003 during the invasion of Iraq, *Operation Telic*. The 120 tanks of 7th Armoured Brigade, part of 1st Armoured Division, went into action around Basra. (UK MoD)

Title page image: An armoured Humvee, manned by US Navy SEALs, pictured in front of Air Force One at Baghdad in late 2003, when President George Bush visited troops in Iraq. (US DoD)

Contents page image: The Humvee was the backbone of US operations in the Iraq war, often being the main form of transport for soldiers. It is still being used today after numerous upgrades. (US DoD)

Back cover image: The Bulldog, formerly the British Army FV 432 armoured personnel carrier, was heavily upgraded to counter roadside bombs and first served on operations at Basra in southern Iraq with the 1st battalion, Royal Green Jackets on *Operation Telic 9* in 2007. (UK MoD)

Published by Key Books
An imprint of Key Publishing Ltd
PO Box 100
Stamford
Lincs PE9 1XQ
www.keypublishing.com

The right of Finlay Reynolds to be identified as the author of this book has been asserted in accordance with the Copyright, Designs and Patents Act 1988 Sections 77 and 78.

Copyright © Finlay Reynolds, 2025

ISBN 978 1 80282 870 2

All rights reserved. Reproduction in whole or in part in any form whatsoever or by any means is strictly prohibited without the prior permission of the Publisher.

Typeset by SJmagic DESIGN SERVICES, India.

Contents

Introduction 4

Chapter 1 United States of America 6

Chapter 2 United Kingdom 45

Chapter 3 Iraq 71

Chapter 4 The Multi-National Force (MNF) 90

Chapter 5 The Legacy of War in Iraq 115

Introduction

Operation *Iraqi Freedom,* the American-led invasion to depose Iraqi leader Saddam Hussein, took place in 2003 and remains the most controversial conflict of modern warfare. It was launched with an assault of US tanks across the open desert from Kuwait to Baghdad, where the 3rd Infantry Division and US Marines faced several days of heavy fighting before they punched a hole through the Republican Guard and secured Baghdad.

The invasion took place after President George W. Bush and his administration announced their intent to remove the dictator from power. As well as tying Saddam and his regime to the events of 9/11, Bush claimed that Iraq had access to weapons of mass destruction and was allegedly poised to use them. Hard evidence was scant and United Nations weapons inspectors failed to find the chemical and biological weapons that the US administration claimed were a danger to the world.

Bush made it clear he wanted to see Saddam removed from power. Although Washington gained support from the UK, there was international condemnation of the plan to invade. Unlike the 1991 conflict, when many Arab nations fully supported the international Coalition and and took part in *Desert Storm*, the military operation to eject Iraq from Kuwait, there was little appetite for war in 2003.

Germany and France voiced their concern about the conditions for going to war and declined to send military forces to participate. Many nations in the Gulf region saw it as a new brand of anti-Arab and anti-Islamic imperialism. Arab leaders objected to the occupation of a fellow Arab country by foreign troops and Washington's intention effectively to secure regime change met with negligible enthusiasm from countries that might otherwise have been relied upon to help. As a consequence, the US Army prepared to deploy a substantial armoured force more or less on its own, with support only from the UK and a small handful of other countries.

This time, more than 350 Abrams tanks as well as sundry armoured vehicles launched a desert assault from Kuwait towards Baghdad, where they planned to capture Saddam and seize the capital. At the same time, more than 150 British Challengers advanced on Basra. In darkness, US and British engineers had breached defensive banks to allow Coalition forces to cross into Iraq on the night of 20 March 2003. The following evening, extensive air and missile strikes hit targets throughout Iraq, in a plan of 'shock and awe' to neutralise Iraqi command infrastructure.

Saddam's military defences soon collapsed in the face of the main US assault. An early objective for the British was to seize the Al-Faw peninsula in southern Iraq, gain access to the vital port of Umm Qasr with the support of Polish Special Forces and occupy what was termed the 'Basra Box'. This was an area of desert that offered protection from Iraqi counterattacks against the initial main northward American advance. After driving towards the Euphrates and making a crossing of the river, US tank units reached Baghdad and moved into the city amid ferocious fighting, which they silenced by 9 April.

The initial US/UK armoured assault saw the destruction of more than 3,000 Iraqi tanks. After the invasion phase, the war gave way to a counter-insurgency operation, and additional protected vehicles were deployed to safeguard soldiers. The assault phase of Operation *Iraqi Freedom* was a military success and demonstrated the importance of tracked armour, but it was regarded by many as a political failure, with ramifications that still resonate in the Middle East today.

Finlay Reynolds

Chapter 1

United States of America

Operation *Iraqi Freedom* – US Forces

The planning and deployment of US forces for Operation *Iraqi Freedom* had started at Bagram Air Base in Afghanistan in 2002, during operations in the Tora Bora mountains. This top-secret headquarters reviewed what would be needed and how the mission would be executed. Planning came after President George W. Bush directed that Saddam Hussein must be removed from power, amid claims that the Iraqi leader had chemical and biological weapons and was ready to use them.

It was soon very clear that this operation would follow in the footsteps of Operation *Desert Storm*, which had ejected Iraqi forces from Kuwait in 1991, and would be spearheaded by Main Battle Tanks

An M1A1 Abrams overlooking Main Supply Route Pluto and Brewers, near Forward Operating Base at Camp Rustamiyah, Baghdad. The tank crew from the 1st Platoon, Delta Company, 2nd Combined Arms Battalion, 69th Armor Regiment was based at Fort Benning, Georgia. This M1A1 is fitted with the Abrams Integrated Management System and the new Tank Urban Survivability Kit (TUSK). (US DoD)

An M1A1 Abrams conducts reconnaissance south of Baghdad in September 2004. The commander and gunner can be seen manning a 50-calibre heavy machine gun and an M240 machine gun. This tank is not fitted with the Tank Urban Survivability Kit, no doubt because it is operating in a rural area. (US DoD)

(MBTs) from the US and its Coalition ally, the UK. They faced a powerful Iraqi army that had built up a significant force of 3,000 tanks and thousands of armoured vehicles.

The thrust across the desert by the 3rd Infantry Division (3ID) and the 1st United States Marine Corps Division (USMC) from Kuwait to Baghdad set a record for an armoured assault, with the lead elements covering 240 miles in just 48 hours to reach the Euphrates crossings at Nasiriyah. Speed was the element that commanders wanted to achieve in order to overwhelm the Iraqi forces before they had a chance to respond. The armoured columns needed constant resupply, so small convoys of fuel bowsers drove to meet up with the 'gas-guzzling' Abrams, the gas-turbine engines of which were particularly fuel-hungry.

As they advanced into Baghdad, the 2nd Brigade's Abrams tanks had to secure Highway 8 from the south; this would eventually become the vital supply line that would sustain the US assault force. Key to that challenge was occupying three intersections, dubbed by army planners in a lighter moment as Objectives 'Moe', 'Larry' and 'Curly', stars of the American comedy *The Three Stooges*. Intense fighting took place at these three locations, each of which was a cloverleaf highway junction of east–west roads with the main north–south route. Successfully holding these highway interchanges was essential to keeping Highway 8 open and allowing US forces to take the city centre following the second Thunder Run, this being a term for a reconnaissance-in-force against remaining Iraqi strength prior to the push into Baghdad. Objective Moe was at the junction of Highway 8 and the Qadisiyah expressway, Larry at Qatar Al-Nada Street leading to the Al Jadriyah bridge, and Curly at the Dora expressway. During the 18-hour battle at Objective Curly, the 3-15 Infantry nearly ran out of fuel and ammunition and was almost overrun but, at the eleventh hour, reinforcements broke through and were able to resupply it. The US force that arrived to help its colleagues included Abrams Main Battle Tanks, Bradley AFVs and Humvees.

M1A1 Abrams tank on patrol in Baghdad. The additional pack can be seen on the side of the tank to protect the crew. This tank is fully closed down, with the driver viewing the road through the armoured windows just below the 120mm main gun. Additional night-vision equipment has been fitted to the turret as well as extra headlights. (US DoD)

As equipment arrived for Operation *Iraqi Freedom*, some tanks were offloaded directly into Kuwait while some Abrams were ferried ashore from amphibious ships by US Navy hovercraft, known as LCACs or Landing Craft Air Cushioned. (US DoD)

Abrams on the outskirts of Ramadi, a town west of Fallujah, which was to become a major battleground as insurgents moved west of Baghdad to focus their attacks on the Coalition and attempt to take control of both Fallujah and Ramadi. Tank operations were almost impossible in these urban areas, where obstacles could be used to slow them down and provide opportunities for the enemy to attack. (US DoD)

Later, as the US forces consolidated in Iraq, examples of the the Mark 9 armoured bulldozer were ferried in by maintenance teams and used to build bases. However, when an insurgent movement arose and began mounting attacks, Mine Resistant Ambush Protected (MRAP) vehicles were introduced to offer troops increased protection. Among the first of these vehicles was the Cougar, a 4x4 with enhanced protection. It was joined by two larger, six-wheeled variants, the Cougar 6x6 and the Caiman. In addition, the Stryker Armoured Personnel Carrier (APC) was deployed by the US military across bases in Iraq. The Stryker, an eight-wheeled armoured personnel carrier, was sent to enhance US capability in Iraq. The Humvee, meanwhile, saw its armour upgraded in late 2003 and several more times subsequently as the campaign evolved.

An Abrams races across the desert in the drive to Baghdad. The M1A1 had a thirsty gas-turbine engine and needed constant re-supply. The lead element of the force reached the outskirts of Baghdad in just 48 hours and then paused as commanders planned the next phase of the battle for the capital. (US DoD)

M1A1 Abrams tanks of the US Marines Regimental Combat Team 1 (RCT1) move to the front of the convoy north of An Nasariyah in central Iraq on 25 March 2003. The Marines pushed north across two routes to fix Iraqi forces in Baghdad and allow 3rd Infantry Division and other Marine units to engage the enemy in the capital. The tank crew had just experienced a huge dust storm that turned the sky orange. (US DoD)

US Marines from 2nd Tank Battalion drive M1A1 Abrams tanks through a sandstorm to northern Iraq in support of Operation *Iraqi Freedom*, 26 March 2003. This unit was part of the 1st Light Armoured Reconnaissance Battalion and forms part of the 1st Marine Division from Camp Pendleton, California. (US DoD)

An M1A1 Abrams tank heads out on a mission from Forward Operating Base MacKenzie, which was part of the Samarra East Air Base, 60 miles north of Baghdad. The Abrams and its crew are assigned to Bravo Troop, 1st Battalion, 4th Cavalry Regiment, 1st Infantry Division. (US DoD)

M1A1 Abrams MBT (Main Battle Tank (MBT) crews from 1st Battalion, 35th Armor Regiment (1–35 Armor), 2nd Brigade Combat Team (BCT), 1st Armoured Division (AD), pose for a photo under the 'Hands of Victory' monument in Ceremony Square, Baghdad. Built at the end of the Iran–Iraq War, this monument marks the entrance to a large parade ground in central Baghdad. The hand and arm are modelled after former dictator Saddam Hussein's own and are surrounded with thousands of Iranian helmets taken from the battlefield. The 24-ton sword blades are cast from the melted guns of dead Iraqi soldiers. (US DoD)

M1A2 Abrams Main Battle Tank

The battle-proven Abrams M1A1 had seen action in the 1991 Gulf War and headed back to the region in 2003 to spearhead Operation *Iraqi Freedom*'s armoured thrust to Baghdad. On 21 March 2003, the US Army's 3rd Division crossed the Kuwaiti border into Iraq and, within days, its Abrams tanks were closing in on Baghdad. Intelligence reports reported negligible Iraqi defences in Baghdad and the commander of the 3rd Division opted to mount an armoured raid into the city. The 2nd Brigade moved up Highway 8 with the aim of linking up with the 1st Brigade at the airport. As the 2nd Brigade advanced, it came under heavy fire but was able to suppress the enemy after intense firefights and reach the airport. In so doing, the Abrams M1A1 had delivered outstanding combat capability in the battle for Baghdad.

As the operation continued into an enduring campaign, the upgraded M1A2 was deployed to Iraq. The M1A2 had an improved commander's weapon station (CWS), plus a commander's independent thermal viewer added to the left side of the turret, forward of the loader's door. These few noticeable external features distinguished the M1A1 and M1A2 from one another; however, the main differences

A pair of US Army M1A2 Abrams pictured in the streets of Baghdad. Both were operating in support of infantry soldiers on the streets. While the commander and gunner can be seen, the driver's position appears to be fully closed down to avoid snipers or blast attack. (US DoD)

lay in new system upgrades and internal adaptations. These included the Inter-Vehicle Information System (IVIS), a secure email system that allowed tank commanders to talk to each other and to their headquarters without the risk of interception by enemy forces. Instead of requiring vehicle crews to monitor the whereabouts and movements of subordinate elements, unit commanders can identify and use data from an onboard Position/Navigation (POSNAV) system. Furthermore, it is possible to locate, map out and distribute enemy positions. Reports and requests for artillery can be automatically written, sent and processed. Finally, the IVIS system allows for the quick distribution of operational orders and can illustrate them with map graphics. The tank's power distribution was enhanced on the M1A2, utilising numerous innovations to ensure that electrical items can still operate through a backup router in the event that one or more conduits are compromised in an attack. The gunner's primary sight has been stabilised in two directions for improved accuracy, while the driver's instrument panel has been updated to a more detailed digital display.

The M1A2 System Enhancement Package (SEP) was a further technology upgrade and standardisation programme that was approved for deployment in 1995. Its goal was to bring the Army's fleet of M1s and M1A2s up to par. The inclusion of an air-conditioning and cooling unit in the crew compartment and the standard under-armour auxiliary power unit are the most noticeable changes. Upgrades to the Gunner's Primary Sight assembly, the tank's radio communications and intercom system, and the IVIS system (incorporating a colour display, full-size keyboard, digital mapping and graphics generation and voice-recognition capabilities) are some of the other changes made to the vehicle.

Following the first batch of 627 units, production of the M1A2 was discontinued. As part of the fleet upgrade programme, the present fleet of M1A2s will undergo a retrofit to bring them up to SEP requirements, while 547 of the Army's inventory of M1s are being updated to M1A2 SEP standards. This will entail the total remanufacturing of the turret.

M1 Abrams Specifications		
Model	M1A1	M1A2
Manufacturer	General Dynamics Land Systems	
Country	United States	
Year	1980 onwards	1993 onwards
Engine	1500 HP Gas Turbine Engine	
Fuel	Multi-fuel	
Protection	Composite Armour	Depleted Uranium
Top Speed	45mph	42mph
Range	265 miles (426km)	
Crew Capacity	4 (commander, driver, gunner, loader)	
Length	32ft (9.77m)	
Width	12ft (3.66m)	
Height	8ft (2.44m)	
Armament	120mm M256 Smoothbore	
Weight	62.8 tons (57 tonnes)	72.7 tons (66 tonnes)
Service Branch	US Army, USMC	

US Army M1A2 Abrams from the 1–3 Combat Regiment in the streets of Al Ba'aj in north-west Iraq. The Abrams crew was supporting a cordon during search operations in the area. Note that towing apparatus has been fitted at the front of the Abrams in case the tank breaks down. (US DoD)

A crew member climbs into an Abrams. The view clearly shows the narrow entrance into the tank and the locking ring for the cupola. The Abrams were used widely across Iraq, with the majority being fitted with enhanced armour. (US DoD)

Abrams tanks patrol central Baghdad. US armour was often deployed to provide close support to foot patrols in the city and reassure the local population that the Coalition was there to deliver security and stability. (US DoD)

US Army M1A2 Abrams with the enhanced Tank Urban Survival Kit (TUSK) fitted. This additional package provided 'explosive reactive armour' to stop rocket-propelled grenades (RPGs) when operating in Baghdad and similar urban environments across Iraq. (US DoD)

Soldiers with the 1452nd Combat Heavy Equipment Transportation Company, North Carolina Army National Guard, load an Abrams tank on to a super heavy equipment transporter tractor-trailer at Contingency Operating Base Adder as part of the US drawdown in Iraq. (US DoD)

M2 Bradley Infantry Fighting Vehicle

The Bradley was among the first armoured vehicles deployed to Kuwait in readiness for Operation *Iraqi Freedom*. After an impressive performance in the 1991 Gulf War, when Bradleys destroyed more Iraqi vehicles than the M1A1 Main Battle Tanks, the M2 formed a vital part of the invasion force this time around. Fast and delivering dynamic firepower, the Bradley was designed to carry a squad of six troops, as well as other specialists in reconnaissance, command and anti-tank roles. A commander, a gunner, and a driver make up the M2's three-person crew, with six fully outfitted troops able to ride along as passengers. The Bradley IFV was designed to be both an armoured personnel carrier (APC) and a tank-killer, in part as a counter to the amphibious Soviet BMP series of infantry combat vehicles. Its design process began in 1963, and manufacturing started in 1981. Its speed had to match that of the new M1 Abrams Main Battle Tank in order to keep formation while moving, something not possible with the older M113 armoured personnel carrier, which was intended to operate in tandem with the older M60. To address various issues that had caused friendly-fire incidents in 1991, infrared identification panels and other marking/identifying systems were externally installed on the Bradleys in 2003.

As the counter-insurgency evolved and extremist attacks soared, the Bradley proved to be susceptible to improvised explosive device (IED) and rocket-propelled grenade (RPG) strikes after 2003, when examples deployed to the urban streets of Mosul, Ramadi, Fallujah and Baghdad.

The M2A3 version of the Bradley started to replace the M3A3 cavalry combat vehicles in US Army armoured reconnaissance units in 2014. This was done because the greater ammunition loads carried by the M3A3s limited the number of scouts (reconnaissance soldiers) that could be transported.

An M2A2 Bradley Fighting Vehicle kicks up plumes of dust in early 2004 as it leaves Forward Operating Base MacKenzie in Iraq. The Bradley was used in both rural and urban environments in support of security operations in Iraq. (US DoD)

Soldiers from the Army's 3rd Armoured Cavalry Regiment load into a Bradley after conducting a combat patrol in the streets of Tal Afar, Iraq, in February 2006. These 30-ton vehicles were fitted with the Bradley Urban Survival Kit (BUSK). (US DoD)

The outbreak of a later international conflict saw 50 Bradleys form part of a $3 billion package of support for Ukraine during the Russian invasion of 2022, as the Pentagon announced on 5 January 2023. France had committed to provide ACMAT Bastions and AMX-10 RCs, with Germany agreeing to send Ukraine examples of the Marder (IFV) as well. US confidence in Ukraine's ability to maintain and sustain such AFVs was key to the supply of Bradleys. Later that month, another bundle of 59 variants was added. Bradleys made their combat debut for Ukraine in mid-April 2023.

Numerous changes have been made to the Bradley series. The M270 Multiple Launch Rocket System (MLRS), the M4 C2V Battlefield Command Post, and the M6 Bradley Air Defence Vehicle, known as the Linebacker (since retired), were all built on its chassis.

Left: A M2A2 Bradley Fighting Vehicle from Apache Troop, 2nd Battalion, 7th Cavalry Regiment, 2nd Brigade Combat Team, 1st Cavalry Division, pictured near Fallujah. Its enhanced armour package can be seen on each side of the vehicle. (US DoD)

Below: US Army soldiers prepare their Bradley Fighting Vehicles at Forward Operating Base MacKenzie in Iraq on 28 October 2004. The enhanced armour can be seen as well as the plates covering the wheels and tracks. (US DoD)

Bradley Infantry Fighting Vehicle Specifications

Model	M2A2	M2A3	M3A3
Manufacturer	United Defence (1981–95) BAE Systems (2004–present)		
Country	United States		
Year	1981–present		
Engine	Cummins VTA-903T 8-cylinder diesel; 600hp		
Fuel	Diesel		
Protection	Spaced laminate armour offering 14.5mm all-round protection. Hull base is 7017 aluminium		Steel, 5083 and 7039 Aluminium
Top Speed	40mph (64km/h) (road) 24.8mph (40km/h) (off-road)		35–41mph (55–66km/h) Road
Range	300 miles (480km)		250–300 miles (400–480km)
Crew Capacity	3 plus 6 passengers		3 plus 2 scouts
Length	21.49ft (6.55m)		
Width	11.82ft (3.60m)		
Height	9.78ft (2.98m)		
Main Armament	25mm M242 chain gun (900 rounds) 2 × TOW anti-tank missile launchers (7 missiles)		25mm M242 chain gun, 1,500 rounds (300 ready) Dual TOW anti-tank missile launcher, 12 rounds (2 in launcher)
Secondary Armament	7.62mm coaxial M240C machine gun (2,200 rounds)		
Weight	27.5 tons (25 tonnes)		25.3–30.8 tons (23–28 tonnes)
Service Branch	US Army		

Soldiers from 1st Battalion, 5th Cavalry Regiment, complete test-firing their 25mm main gun and M240B machine gun during a test-firing exercise near Operating Base Speicher in Iraq. (US DoD)

A US Army Bradley crew of Charlie Troop, 1st Battalion, 4th Cavalry Regiment, 1st Infantry Division, fire their Bradley's 25mm chain gun during operations near their base at Ad-Dawr, Iraq, in November 2004. It was here that Saddam Hussein was captured in December 2003. (US DoD)

A US Army Bradley on the outskirts of Baghdad after engaging an Iraqi BMP-2 armoured vehicle, which is burning among the trees. The Bradley distinguished itself in Iraq, defeating several T-72 tanks. (US DoD)

Humvee – High Mobility Multipurpose Wheeled Vehicle

The iconic Humvee (High Mobility Multipurpose Wheeled Vehicle or HMMWV) was deployed in *Iraqi Freedom* in various roles including escort, reconnaissance and as a command platform, almost 20 years after it was first introduced. It could carry four soldiers and travel 300 miles across undulating terrain and had a reputation as a robust 'battle wagon'.

During the advance to Baghdad, anti-tank teams travelled in Humvees behind the Abrams to defend the tanks. They were known as Combined Anti-Armor Teams (CAAT) Teams and were used with great effect by the US Marines. These units, often in groups of four or five Humvees, were equipped with the BGM71 TOW (Tube-launched Optically Tracked wire-guided weapon) and drove directly behind the tanks or on the flank ready to protect them. Humvees were also used by Tactical Air Control Parties (TACPs), regimental planning teams, medical units and engineers.

After the initial push into Iraq in 2003, the mission changed. Once in Baghdad, the Humvee was no longer responsible for whisking soldiers and Marines across wide expanses of sand. Instead, it suddenly became a combat taxi in the world's most dangerous country. As insurgent attacks left the Humvee vulnerable, plans were quickly drawn up to add additional armour to the vehicle. The Armour Survivability Kit (ASK) was introduced in October 2003, adding about 1,000lb (450kg) to the Humvee's weight. In 2006, the improved Fragmentation Kit 5 was introduced. Its Frag 5 armour upgrade included four 600lb doors with additional plating of homogeneous steel armour as well as a large armoured cupola to protect the gunner from attack. Armoured doors and windows were fitted, plus protective plates under the chassis. Most up-armoured variants held up well against attacks, but as

Humvees advance across the desert towards Baghdad during a sandstorm. These vehicles could carry four soldiers in what were cramped conditions. The leading vehicle is fitted with an anti-tank weapon but at this point Humvees were not fitted with the extra armour packs. (DPL)

the armour increased, the insurgents used correspondingly bigger roadside bombs. In 2008, the Frag Kit 6 armoured upgrade was phased in. It had been developed by the US Army Research Laboratory to defeat explosively formed projectiles (EFP), a type of armour penetrator often utilised in IEDs. The Frag Kit 6 added 1,000lb of extra weight and increased the width by 12in. Humvees fitted with this kit were known as the M1151.

Humvee – High Mobility Multipurpose Wheeled Vehicle (HMMWV) Specification	
Model	M1114 UAH | A later fully armoured variant was listed as M1151
Manufacturer	AM General
Country	United States
Year	1984–present
Engine	6.2 L V8 diesel or 5.7L gasoline or 6.5L V8 turbo diesel and non-turbo diesel: 190hp
Fuel	Diesel
Protection	Welded aluminium, composite and steel armour protection (varies on armour package)
Top Speed	55mph at max gross weight. Over 70mph top speed
Range	300 miles (480km)
Crew Capacity	4; varies on configuration. (2 in front, 2 in rear)
Length	16ft (4.8m)
Width	7ft – increased by 12in on the M1151 (2.1m)
Height	6ft (1.82m)
Main Armament	Multiple configurations. Typically, M2 HMG, M134 minigun or MK19 grenade launcher. Can also carry anti-tank system.
Weight	6.61 tons (6 tonnes)
Service Branch	US Army

Humvees patrol through the streets of Baghdad shortly after the invasion in 2003. They have been fitted with an armoured cupola for the gunner on top of the vehicle, providing protection against blast bombs. (DPL)

US Army Humvee of the 3rd Infantry Division on patrol with a Bradley AFV in Baghdad. It is fitted with an interim armoured shield providing front and rear protection for the gunner. (DPL)

Humvees of the 101st Airborne, fitted with heavy machine guns and anti-tank weapons. The weather was so hot that soldiers removed the vehicles' doors, but as insurgent attacks soared, armoured replacements were quickly re-installed when the vehicles were upgraded. (US DoD)

Above: Armoured Humvees line up for an operational deployment. An armoured cupola has been fitted along with doors and front panels. The thickness of the blast-proof doors significantly altered the vehicle's profile. (DPL)

Left: An armour-upgraded Humvee, which protected soldiers from IED blast and small-arms fire. The heavy doors and the huge cupola on top of the vehicle added significant weight, while the platform was also fitted with run-flat tyres. (US DoD)

MRAP (Mine Resistant Ambush Protected) Vehicles

By late 2003, the insurgency in Iraq had grown and was mounting attacks on a daily occurrence. US commanders identified a need for 'better protected' vehicles that could withstand the threat posed by improvised explosive devices (IEDs) as well as conventional mines and ambush tactics used against Coalition vehicles. A new range of vehicles was designed, known as MRAPs (Mine Resistant Ambush Protected) vehicles. They included the Cougar, which was produced by the US company, Force Protection. It was ordered in 2004 and examples were delivered for operational use in 2007. It had two air-conditioning units, NBC overpressure and filter protection and an electrically-powered winch, capable of hauling a four-ton capacity. The Cougar was produced with run-flat tyre inserts and was able to travel both on and off road. It could also be transported by C-17. Most importantly, the Cougar-H armour kit offered complete all-round protection against 7.62mm rounds (the glass fitted was already resistant to multiple 7.62mm strikes). During trials, a 30lb TNT bomb placed between the front and rear axles and a 15lb charge placed under the centre of the chassis were detonated without harming the Cougar-H. The radiator, tyres, battery compartment, gasoline tanks, engine and gearbox are all covered by ballistic protection. The V-shaped hull's purpose is to deflect the blast outside and away from the car's passenger cabin. The explosion may render the vehicle unusable, but the test concluded that the occupants would not be hurt. Other variants included the Cougar HE (6x6 wheeled version) and the Caiman (6x6).

The MRAP class (all-wheeled) is separated into three categories according to weight, size and purpose. Category I includes 4x4 vehicles weighing about 7 tons and designed for use in urban environments. These vehicles can transport up to six personnel. Category II (MRAP-JERRV) is based on 6x6 vehicles

US Marines from the 1st Explosive Ordnance Disposal Company, II Marine Expeditionary Force, prepare to board a four-wheel drive Cougar at Al-Taqaddum Air Base in Al Anbar province. (US DoD)

weighing an estimated 19 tons. These are mainly used for convoy escort, troop transport and ambulance evacuation, transporting groups of up to ten personnel. Category III vehicles are those intended to be used primarily on route clearance and explosive ordnance disposal, weighing about 22.5 tons and capable of carrying up to 12 passengers.

MRAP (Mine Resistant Ambush Protected) Specification			
Model	Cougar H (4x4)	Cougar HE (6x6 Variant)	Caiman 6x6
Manufacturer	General Dynamics Land Systems (Formerly, Force Protection Inc)		BAE
Country	United States		
Year	2002 (produced); in service 2004–07		2007–present
Engine	Caterpillar C7 diesel 330hp		Caterpillar C-9 diesel 450hp
Fuel	Diesel		
Protection	Classified		
Top Speed	65mph		68mph
Range	600 miles (966km)		401 miles
Crew Capacity	2 plus 4 passengers	2 plus 8 passengers	2 plus 8 passengers
Length	19.41ft (5.91m)	23.25ft (7.08m)	25.35ft (7.72m)
Width	9ft (2.73m)		8.1ft (2.47m)
Height	8.67ft (2.64m)		9.24ft (3.09m)
Main Armament	M240 7.62mm machine gun or M2 .50 cal machine gun or optional remote weapons system		
Weight	15.9 tons (14.5 tonnes)	18.9 tons (17.2 tonnes)	18.5 tons (16.8 tonnes)
Service Branch	USMC, US Army, USN		

A US Marine mans an M2 .50-calibre heavy machine gun during crew-served weapons training. MRAPs could be equipped with a variety of different weapons. (US DoD)

A six-wheeled Cougar. Designed with a heavily armoured monocoque V-shaped hull to protect against insurgent attacks, the vehicle was built to withstand landmines and IEDs. (US DoD)

During trials, the 6x6 Cougar was subjected to extensive explosive testing to ensure it was capable of withstanding close-range detonations from IEDs and roadside bombs in Iraq. (US DoD)

American soldiers serving with Delta Company, 1st Battalion, 186th Infantry Regiment of the 41st Infantry Brigade Combat Team head out from Camp Adder in southern Iraq, aboard six-wheeled Caiman mine-resistant vehicles. A M1117 Guardian armoured security vehicle can be seen in support. (US DoD)

A new Caiman driver serving with the 9th Cavalry Regiment of the 3rd Advise and Assist Brigade, is put through his paces at Operating Base Delta. The Caiman is fitted with an enhanced armour package, armoured cupola and has a V-shaped hull design to deflect mines and roadside bomb blasts. (US DoD)

US soldiers from the 2nd Brigade Combat Team of the 4th Infantry Division arrive by Black Hawk helicopters and Caiman protected vehicles to assist the Iraqi Army in distributing humanitarian aid to the citizens of Faddaqhryah and Bahar in the Basra Province of southern Iraq. (US DoD)

M1126 Stryker Combat Vehicle

The eight-wheeled 20-ton armoured Stryker combat vehicle was deployed to Iraq after the main invasion and was used as a battlefield taxi, providing protected travel for soldiers on patrol in the country's major cities. When operating in urban areas, the Stryker crew would deploy its soldiers on patrol and then follow in support with its heavy machine guns, providing covering protection if needed. In development, the Stryker Interim Armoured Vehicle (IAV) project offered the US Army a family of ten distinct vehicles built on a single chassis. This provided frontline units with a range of combat capabilities, based on two types of Stryker platforms; the Mobile Gun System (MGS) and the Infantry Carrier Vehicle (ICV). The ICV is available in eight more configurations: the Reconnaissance

Above: A US Army Stryker assigned to the 172nd Stryker Brigade Combat Team monitors activity on a main road in northern Iraq. The US Army used this vehicle on operations for the first time in Operation *Iraqi Freedom*. (US DoD)

Right: A Stryker team of the 2nd Battalion, 11th Field Artillery Regiment serving with the Multi-National Division – Baghdad escorts members of the regional Provincial Reconstruction Team at Sab al Bour, north-west of Baghdad. (US DoD)

Vehicle (RV), Medical Evacuation Vehicle (MEV), Engineer Squad Vehicle (ESV), Commander's Vehicle (CV), Fire Support Vehicle (FSV), Mortar Carrier (MC), Anti-Tank Guided Missile Vehicle (ATGM) and NBC Reconnaissance Vehicle (NBCRV). The first systems were delivered to the Army from General Dynamics Land Systems in February 2002. The MGS and NBCRV were delivered beginning in 2004 and served extensively in Iraq. It remains in service and is seen as a key element of the modern infantry.

M1126 Stryker Specification	
Model	MGS/ICV
Manufacturer	General Dynamics Land Systems
Country	United States
Year	2002–present
Engine	Caterpillar C7 350hp
Fuel	Diesel
Protection	14.5 × 114mm protection with ceramic bolt on armour
Top Speed	60mph
Range	310 miles (480km)
Crew Capacity	2 but varies. Up to 9 passengers
Length	22.10ft (6.95m)
Width	8.11ft (2.72m)
Height	8.8ft (2.64m)
Main Armament	Configurations include: M2 machine gun, Mk19 grenade launcher, Mk44 Bushmaster gun, M68A2 gun
Weight	18.1 tons (16.47 tonnes) (ICV Variant) 20.6 tons (18.77 tonnes) (MGS Variant)
Service Branch	US Army

A patrol moves through the streets of Sab al Bour, north of Baghdad, escorted by Stryker armoured vehicles. The Strykers, from the 104th Cavalry Regiment, 56th Stryker Brigade Combat Team, are fitted with plate armour and cages to stop rocket-propelled grenades. (US DoD)

Inside the rear of the Stryker, soldiers sit facing each other. A monitor allows the team to view the environment outside, while hatches in the rear roof allow soldiers to observe while remaining inside the protected vehicle. (US DoD)

The Stryker remains in service and has been upgraded several times to incorporate a range of surveillance systems. These vehicles in service with the 2nd Cavalry Regiment are among the most recently upgraded. (Fin Reynolds/DPL)

A AV7A1 AMTRAC – Amphibious Armoured Personnel Carrier

The AAV7A1 is an amphibious armoured personnel carrier, often known as the Amtrac. It entered service in 1972 as a replacement for the ageing LVTP-5, which had been in service with the US Marines since the 1950s. At first known as the LVTP7, its designation was changed by the Marine Corps in 1984 to AAV7A1 (amphibious assault vehicle), representing a shift in emphasis away from the long-time LVT (amphibious) designation, meaning 'landing vehicle, tracked'. Without a change of a bolt or plate, the AAV7A1 was to be more of an armoured personnel carrier and less of a landing vehicle. Weighing in at 26 tons combat-loaded, and with a three-man crew, it can officially carry 21 personnel. With a road speed of 45mph, this vehicle is fully amphibious, with water speeds up to 8mph. It is not as heavily armoured as the US Army's Bradley Infantry Fighting Vehicle.

Left: Wearing protective chemical clothing, Marines from the 15th Marine Expeditionary Unit, supported by an Amtrac amphibious armoured carrier, move towards an objective in Az Zubayr, Iraq, on 23 March 2003. (US Marines)

Below: Amphibious Armoured Assault Vehicles belonging to Charlie Company, 1st Battalion, 5th Marines, Regimental Combat Team 5, 1st Marine Division, move along an Iraqi highway during a sandstorm on 24 March 2003. (US Marines)

United States of America

In November 1990, a large-scale amphibious exercise, with codename *Imminent Thunder*, was held near the head of the Persian Gulf. Although the surface assault was cancelled, the concept created the basis of a deception plan, which left Iraqi commanders thinking the Coalition would strike from the sea – they didn't. The USMC deployed several Amtrac variants with success in *Desert Storm* during the liberation of Kuwait and destruction of Saddam's forces, although they did prove vulnerable to Iraqi rocket attacks.

AAV7A1 AMTRAC (US Marines: renamed from LVTP7 in 1984 by USMC) Specification	
Model	Amphibious Armoured Personnel Carrier
Manufacturer	FMC Systems
Country	United States
Year	1972–present
Engine	Detroit Diesel/Cummins
Fuel	Diesel
Protection	45mm (1.8in) of armour plate
Top Speed	20mph off-road; 45mph on surface road (72km/h), 8mph in water
Range	300 miles (480km) – 20 nautical miles in water at sea state 5
Crew Capacity	3 crew (commander, gunner and driver) plus 21 passengers
Length	26.1ft (7.94m)
Width	10.9ft (3.27m)
Height	10.8ft (3.26m)
Armament	Mk19 40mm automatic grenade launcher – 96 rounds ready and 768 stowed; 12.7mm M2HB heavy machine gun, 200 rounds ready, 1,000 stowed
Weight	32 tons (29.1 tonnes)
Service Branch	United States Marine Corps (USMC)

Amphibious Armoured Assault Vehicles cross a river in Iraq during Operation *Iraqi Freedom*. The Amtrac proved its capability in providing armoured protection in the amphibious role. (US Marines)

Marines assigned to Bravo Company, 2nd Battalion, 2nd Marine Division, use amphibious armoured assault vehicles to block a roadway at Amiriyah in Fallujah during search operations. (US Marines)

An Amphibious Assault Vehicle drives through a wall and locked gate to open a path for Marines of India Company, 3rd Battalion, 1st Marines, on offensive operations at Fallujah in Iraq. (US Marines)

Above: US Marines serving with Company D of the Light Armoured Reconnaissance Battalion Landing Team 3/1 await orders to move forward during a patrol south of Baghdad. (US DoD)

Right: A Light Armoured Reconnaissance Vehicle (LAV) provides security for the 1st Light Armoured Reconnaissance Battalion (LAR) outside Fallujah, Iraq. It was in Fallujah that the Marines saw extensive combat against insurgents. (US DoD)

Below: A US Marines LAV-25 patrol pauses on a hill overlooking the city of Hit in Iraq. The crew hang items of their equipment and supplies on the outside of the vehicle to give themselves more space inside. Behind the cupola, fuel can be seen stowed, while on the side, bergens (backpacks) hang from fixed points. (US DoD)

LAV-25 – Armoured Light Reconnaissance Vehicle (USMC)

The LAV-25 was deployed in the initial phase of Operation *Iraqi Freedom* by the US Marine Corps (USMC). This eight-wheeled light armoured reconnaissance vehicle, its design based on the Swiss Piranha series of wheeled vehicles, was developed in the late 1980s after the USMC sought a light armoured vehicle (LAV) that could deliver speed and protection. In total, 758 of what was termed the LAV-25 were ordered, being delivered in command, anti-tank, recovery and logistics roles, and their first use in combat was in the invasion of Panama in 1989, before subsequent deployment to the Persian Gulf.

It was in that conflict that the all-terrain capability of the LAV-25 appealed to the 82nd Airborne, which borrowed at least a dozen and deployed them in combat. The LAV-25 can perform a range of tasks due to its speed and manoeuvre, while delivering suppressive fire against armoured and thin-skinned vehicles with its 25mm chain gun. Its amphibious capability enables crossing of inland waterways, rivers and streams with the least amount of preparation, as the vehicles can be in the water within three minutes. Depending on the mission requirements, additional weapons can also be carried, such as anti-tank missiles or machine guns. By 2003, the LAV-25 was seen as a workhorse of the USMC.

LAV-25 Specification (US Marines) Specification	
Model	Armoured Reconnaissance Vehicle
Manufacturer	General Motors | General Dynamics Land Systems
Country	United States/Canada
Year	1993–present
Engine	Detroit Diesel 6V53T 300hp
Fuel	Diesel
Protection	Welded steel
Top Speed	62mph (100km/h), 6mph (9km/h) in water
Range	410 miles (659km)
Crew Capacity	3 crew (commander, gunner and driver) plus 6 passengers
Length	21ft (6.39m)
Width	8.2ft (2.5m)
Height	8.10ft (2.69m)
Armament	M242 Bushmaster 25mm chain gun plus two 7.62mm machine guns
Weight	14.1 tons (12.8 tonnes)
Service Branch	United States Marine Corps

M113 – Armoured Personnel Carrier

The M113 Armoured Personnel Carrier (APC) was deployed as part of Operation *Iraqi Freedom* in 2003, serving in northern Iraq with Task Force 1–63 and with the 3rd Infantry Division in its drive to Baghdad.

The US Army began working on the development of new armoured personnel carriers (APCs) after World War Two. Designed by the FMC Corporation, the M113 was a fully tracked APC and first entered service in 1961 to replace the M59, which had been in operation since the early 1950s. It was officially replaced with the Bradley APC, but more than a dozen variants of the versatile and highly reliable M113 remained in service. These included the M163 Vulcan Air Defence System, which was fitted to the hull of an M113 and consisted of a Vulcan Gatling gun, plus the M901 ITV (improved TOW anti-tank weapon), a command vehicle and an ambulance variant.

By 2003, the M113 had received bolt-on armour, which added stress to the engine and gearbox. Even so, this veteran APC proved its value in the drive to Baghdad. As they joined logistics convoys on the heels of the armoured advance, US Army engineer vehicles with the 3rd Infantry Division came under fire from small arms and RPGs. An after-action inspection revealed that the trailers had had their wheels shot out, and the trailers themselves were riddled with bullet holes. The M113s had dragged these explosives-packed trailers through a fire zone.

The US Army stopped buying M113s in 2007, with almost 6,000 vehicles remaining in its inventory. The M113 will be replaced in US Army service by the Armoured Multi-Purpose Vehicle (AMPV) project.

M113 Armoured Personnel Carrier: Specification	
Model	Armoured Personnel Carrier
Manufacturer	General Motors | General Dynamics Land Systems
Country	United States
Year	1960–present
Engine	Detroit diesel
Fuel	Diesel
Protection	5083 aluminium armour 8–44mm
Top Speed	42mph (67km/h), 4mph (6km/h) in water
Range	300 miles (482km)
Crew Capacity	2 (commander, driver) plus 11–15 passengers
Length	15.11ft
Width	8.97ft
Height	8.2ft
Armament	M2 Browning machine gun
Weight	13.5 tons (12.3 tonnes)
Service Branch	United States Army

US Army M113 armoured personnel carriers deployed into Iraq in March 2003 with their standard armour. As insurgents mounted their offensive against the US-led Coalition, armoured protection was increased. (US DoD)

Above: An American M113 with enhanced armour and metal grilles to protect from rocket-propelled grenades as well as being fitted with an armoured protected cupola. The M113 was quickly replaced with the MRAP. (US DoD)

Left: US soldiers prepare to fire heavy mortars from an M113 that has been upgraded to protect from IEDs and rocket-propelled grenades (RPGs). This vehicle has one of the early cupolas designs to protect the gunner. (US DoD)

M60 – Armoured Vehicle Launched Bridge (AVLB)

The M60 tank entered service in 1959 and was used in the Gulf War of 1991 by the US Marines. By 2003, it had been retired, but the chassis was used as the platform for the Armoured Vehicle Launched Bridge (AVLB). This combat engineer vehicle was developed by the US Army Engineer Research and Development Laboratories in 1963 and was one of the oldest vehicles deployed by the US Army. It was designed to launch bridges to take tanks and other wheeled combat vehicles across trenches and water obstacles in combat conditions. A total of 400 armoured bridge launchers and bridges were built. The AVLB had no weapon platforms and was assigned protection when deployed in the frontline. In the war of 2003, the AVLB was an important piece of equipment for the advancing task force.

M60 AVLB Specification	
Model	M60 | AVLB
Manufacturer	Chrysler Corporation Delaware Defense Plant
Country	United States
Year	1960, subsequently upgraded in 1962, 1974 and 1979
Engine	Continental AVDS-1790 series turbo-supercharged, fuel injection 12-cylinder 750hp (559kw)
Fuel	Diesel
Protection	Limited – specialist bridge-laying capability
Top Speed	30mph (48km/h)
Range	310 miles (498km)
Crew Capacity	2
Length	31ft (9.44m)
Width	12ft (3.66m)
Height	10.10ft (3.3m)
Armament	None
Weight	56.2 tons (51 tonnes)
Service Branch	US Army

The US Marine crew of an M60A1 Armoured Vehicle Landing Bridge (AVLB) practise the deployment of its 60ft-bridge span in Kuwait just before the invasion of Iraq. It was designed to quickly allow heavy wheeled and tracked military vehicles to move over unstable or hazardous terrain. (US DoD)

US Marine Corps (USMC) personnel from Charlie Company, 1st Tank Battalion, drive an M60A1 Armoured Vehicle Launched Bridge (M60A1 AVLB) along route Tampa towards a forward operating base near Jalibah Airfield, during the early days of Operation *Iraqi Freedom*. (US DoD)

D9 Armoured Combat Earthmover (ACE)

The D9 Armoured Combat Earthmover (ACE) is a highly mobile armoured tracked vehicle that provides combat engineering support to frontline forces. Fielded by both the United States Marine Corps and US Army, its tasks include eliminating enemy obstacles and carrying out maintenance and repair of roads and supply routes, as well as the construction of forward operating bases. The engine, drive train and driver's compartment are situated towards the rear of the vehicle, while the front comprises a dozer blade with a composite aluminium ejector that can unload ballast and or cargo quickly in combat or hostile conditions. The D9 is often in the vanguard of the frontline, clearing obstacles or, in the case of Operation *Iraqi Freedom* of 2003, cutting a path through sand berms – huge walls of sand built to protect Coalition forces. Its exposed nature can be offset through the fitting of additional Explosive Reactive Armour (ERA) plates.

The main vehicle hull is produced from welded and bolted aluminium with a two-speed winch capable of pulling 25,000lb. Towing pintle and airbrake connections are provided. It is equipped with a suspension system that allows the front of the vehicle to be raised, lowered, or tilted to permit dozing, excavating, rough grading and ditching functions. The D9 is armoured against artillery fragmentation and, with the additional armour can withstand rocket attacks and roadside bombs. It has smoke screening capability and has chemical-biological protection for the operator. Road speed is 30mph (48km/h). It is transportable in a C-130 and C-17. Prior to Operation *Iraqi Freedom*, the United States purchased tractor protection kits from Israel for Caterpillar D7 bulldozers, which the US Army deployed to operate alongside the D9. The armoured bulldozers were mainly used in mine-clearing applications. They were heavily fortified and were used to clear destroyed vehicles from roads, plus dig moats and erect earthen barriers.

The D9 Armoured Combat Earthmover (ACE) was a highly mobile armoured tracked vehicle that provided combat engineering support to frontline forces in Iraq. Fielded by both the United States Marine Corps and US Army, its tasks included eliminating enemy obstacles and maintenance and repair of roads. (US DoD)

D9 Armoured Earth Mover

Model	D9 Armoured Earth Mover
Manufacturer	Caterpillar and International Harvester
Country	United States
Year	1986 onwards
Engine	Cummins V903C 8-cylinder
Fuel	Diesel
Protection	Classified
Top Speed	30mph (48km/h)
Range	200 miles (322km)
Crew Capacity	1
Length	20.50ft (6.25m)
Width	10.49ft (3.2m)
Height	8.8ft (2.7m)
Armament	Crew personal weapons
Weight	27.5 tons (25 tonnes)
Service Branch	US Army, USMC

The D9 is often in the vanguard of the frontline, clearing obstacles or, in the case of Operation *Iraqi Freedom*, **cutting a path through sand berms. It can be fitted with additional Explosive Reactive Armour(ERA) plates. (US DoD)**

D7 Armoured Caterpillar

The Caterpillar D7 is a medium track-type tractor manufactured by Caterpillar and used mainly by the US Army as a bulldozer. In Iraq, the D7 was deployed to build huge protective walls of sand, called berms, to hide and protect Coalition forces. These powerful vehicles were a key part of the combat engineering equipment deployed in Operation *Iraqi Freedom*. As the invasion force stood ready to advance into Iraq, the D7s were sent ahead to cut a path through the sand walls that had been built to shield the assembling forces. Manufactured since 1938, the D7 has seen numerous mechanical improvements, but it is slow and an easy target for enemy forces. As a result, when operating in the frontline, it is routinely fitted with enhanced armour packages. These, which included a complete armoured shield for the driver, were procured from Israel to protect the crewmen from attack as well as when clearing obstacles and using the dozer blade to make defences around forward operating bases. A towing hook and heavy-duty winch are fitted to the rear along with an entrenching tool, which allows the D7 to dig a cable trench for communications. This unique item of armoured equipment remains in service today.

Above: The Caterpillar D7 is a medium-track-type tractor used by the US Army that was heavily armoured for its frontline role in Iraq. These powerful vehicles were a key part of the combat engineering equipment deployed on Operation *Iraqi Freedom*, clearing obstacles and building berms. (DPL)

Opposite: Soldiers from Alpha Company, 3rd Special Troops Battalion of the 101st Airborne Division, prepare to unload two armoured Caterpillar D7 bulldozers from their carriers at Al Butoma in Iraq. The soldiers will use the bulldozers to construct sand berms for security in Al Butoma. (US DoD)

D7 Armoured Caterpillar Bulldozer Specification

Model	D7 Bulldozer
Manufacturer	Caterpillar
Country	United States
Year	1938 onwards
Engine	Cummins C9.3 diesel engine
Fuel	Diesel; 2009 versions are fitted with a diesel-electric drive
Protection	Steel armour, fitted by Israeli Military Industries
Top Speed	15mph (24km/h)
Range	200 miles (322km)
Crew Capacity	1 – sometimes 2
Length	13.45ft (4.09m)
Width	8.2ft (2.5m)
Height	7.87ft (2.4m)
Armament	Crew personal weapons
Weight	15.8 tons (14.4 tonnes)
Service Branch	US Army, USMC

LAV-25 Light Armoured Vehicle

The LAV-25 is an armoured amphibious personnel carrier introduced in 1983 to provide greater protected mobility for the USMC. The 14-ton vehicle is crewed by three and can carry six soldiers, and its Detroit Diesel engine allows the eight-wheeler to deliver an operational range of 410 miles with a maximum speed of 62mph.

During the 1991 Gulf War, the US Army borrowed a number of LAV-25s from the USMC. In January 2003, the 1st Light Armoured Reconnaissance Battalion was deployed to Kuwait in readiness for the invasion of Iraq. The battalion was the first unit to cross the Iraqi–Kuwaiti border on 20 March. After driving northward through Iraq, this unit was instrumental in securing Baghdad. Due to its unique mobility and reconnaissance capabilities, the battalion then left Regimental Combat Team Five and was assigned to Task Force Tripoli, formed to take Tikrit, Saddam Hussein's home town north of Baghdad. After the cessation of major combat operations, the 1st Light Armoured Reconnaissance Battalion moved from the northern portion of Iraq to the country's extreme southern area along the Saudi-Arabian border to halt and deter illegal smuggling into Iraq. It then redeployed back to the United States at the end of May 2003.

LAV-25 Specification	
Model	Light Armoured Reconnaissance Vehicle
Manufacturer	General Motors | General Dynamics Land Systems
Country	United States/Canada
Year	1993–present
Engine	Detroit Diesel 6V53T 300hp
Fuel	Diesel
Protection	Welded steel
Top Speed	62mph (100km/h), 6mph in water
Range	410 miles (660km)
Crew Capacity	3 (commander, gunner and driver) plus 6 passengers
Length	21ft (6.39m)
Width	8.2ft (2.50m)
Height	8.10ft (2.69m)
Armament	M242 Bushmaster 25mm chain gun plus two 7.62mm machine guns
Weight	14.1 tons (12.80 tonnes)
Service Branch	United States Marine Corps. Twelve loaned to 82nd AB, US Army

Chapter 2
United Kingdom

Operation *Telic* – British Forces

As the countdown to war approached in 2003, Britain's Prime Minister, Tony Blair, offered US President George W. Bush his support in removing Saddam Hussein from power in Iraq. The Bush administration claimed that Saddam was actively developing biological and chemical armaments, so-called weapons of mass destruction (WMDs), in contravention of international treaties and protocols. However, Blair's plan to send the UK's armed forces was delayed, as many of his own cabinet rebelled against the government's choice to go to war alongside the US. Several of Blair's ministers stood down, including his foreign secretary, Robin Cook, who made his objections clear in doing so. Millions took to the streets in the UK and across Europe, marching in protest in an attempt to stop the war. Then, in January 2003, Blair and his Cabinet defeated a House of Commons anti-war vote with a majority of 179. While this late go-ahead was significant for the government, it delayed the deployment of some military equipment, including body armour.

The preferred option of Britain's military chiefs was to use Turkish bases and attack Iraq from the north, through friendly Kurdish territory. However, Ankara declined to support the war and a decision was eventually made to attack from the south, using Kuwait as a launch pad. The movement of British troops and equipment to Kuwait began in earnest in January 2003. This represented a tremendous logistical challenge for the three branches of the armed services, which encountered significant shortages and delays in supplying troops with the necessary equipment in time, such as boots and NBC clothing. Large tented camps were constructed in Kuwait as the force commenced its operational readiness training. However, widespread dust storms delayed progress. In addition, the potential threat from biological and chemical attacks caused commanders to take extra precautions.

On 20 March 2003, the British headed for Basra, Iraq's second city, as the US moved towards Baghdad – the invasion of Iraq had begun. The Royal Scots Dragoon Guards (SDG), the Queen's Royal Lancers (QRL) and the Royal Tank Regiment (RTR) deployed Challenger tanks, which spearheaded the advance into Basra. Elements of the SDG supported the Royal Marines in the Al-Faw peninsula, while the QRL and RTR moved into the streets of Basra. The battle for Basra lasted from 21 March to 6 April and was one of the first major engagements of the invasion. The 7th Armoured Brigade fought its way into the city against constant attack from the Iraqi Army's 51st Division and the Fedayeen Saddam, meaning 'Saddam's men of sacrifice'. This was a fanatically loyal unit within the Ba'ath party, with almost 25,000 fighters in Basra. As an independent military unit, the Fedayeen Saddam had a fierce reputation, but when Basra fell, most of its fighters fled.

On 14 April, Blair announced to the House of Commons that Saddam's regime had been removed and that the bulk of Iraq was under Coalition control. Armour had been vital to the initial success of the operation, with Challenger tanks able to quickly advance to the outskirts of Basra, support infantry forces and send a clear signal of military power to the Iraqi people, while Warriors, Scimitars, Bulldogs and MRAPs all made a significant contribution to force protection.

After the overthrow of Saddam Hussein's regime, Iraq was occupied, though the country was riven by an insurgency for a further six years.

FV4034 Challenger 2 Main Battle Tank

In March 2003, the British Army's Challenger 2 tanks were loaded on to ships at Marchwood in Hampshire and headed for the Middle East. The British force that deployed to Kuwait included 120 Challengers from the Royal Scots Dragoon Guards, the 2nd Royal Tank Regiment and the Queen's Royal Lancers. They were part of the 7th Armoured Brigade, known as the Desert Rats, and headed the advance into Basra, the main city in the south of Iraq during Operation *Telic*.

The Challengers suffered no losses to Iraqi fire. In one encounter within an urban area, a Challenger 2 came under attack from irregular forces with heavy machine guns and rocket-propelled grenades. The driver's sight, directly under the main gun, was damaged and, while the tank attempted to back away under the commander's directions, the other sights were also damaged and the tank lost a track as it reversed into a ditch. The crew survived, safe within the tank until it was recovered for repairs, the worst damage being to the sighting system. It was back in operation six hours later.

In an incident on 27 March, 14 Challenger 2 tanks of the Royal Scots Dragoon Guards destroyed a squadron of ageing Iraqi T-55 tanks and three armoured personnel carriers. The Challengers had been summoned to support Royal Marine Commandos securing the Al-Faw peninsula further south of Basra. As they crossed a pontoon bridge, built by the Royal Engineers, 15 Iraqi tanks emerged from a wooded area and engaged the Challengers. In a fierce firefight, all the T-55s were destroyed. In the same week Sgt Steven Roberts, a tank commander with the Royal Tank Regiment, was killed in an incident near Basra. It was later revealed at an inquest that there was a shortage of body armour and the sergeant had given his own body armour away to another soldier.

The Challenger 2 had undergone significant upgrades for its deployment in the desert, which included being fitted with the highly secret Chobham armour and spaced ceramic plates for exceptional

A Challenger 2 Main Battle Tank of the Queens Royal Lancers crosses an Iraqi defensive ditch during the early week of Operation *Telic*, the British mission during the invasion of Iraq. Chobham armour can be seen on the right side of the tank. (UK MoD)

protection. Its 120mm smoothbore gun gave the enabled crew to fire various types of ammunition and engage enemy targets at distance. As well as the Challenger Main Battle Tank, the UK sent Warrior armoured personnel carriers and much older troop carriers called FV432s, which supported the Challengers.

The Challenger 2, otherwise known as CR2, is a third-generation main battle tank designed by Vickers (now owned by BAe Systems). It was an upgrade of the Challenger 1, which had been deployed to the Middle East in 1991 to take part in Operation *Granby*, the British participation in the US-led Operation *Desert Storm*. The Challenger 2 has four crew members, consisting of a commander, gunner, loader and driver. The platform's turret and hull are protected with second-generation Chobham armour, also known as Dorchester. The tank is powered by a Perkins CV12-6A V12 diesel engine and has a range of 340 miles with a maximum road speed of 37mph.

After the Iraq War, the tank saw further improvements, including the installation of the Bowman tactical secure digital communications system with a built-in GPS receiver, which itself has since been upgraded. The British Army is currently investing in a new Challenger 3, which will provide the military with 148 tanks. In May 2024, successful firing tests were conducted in Germany, highlighting the Challenger 3's exceptional long-range capabilities. Central to these tests was the tank's cutting-edge Rheinmetall L55A1 120mm L55A1 high-pressure smoothbore gun, paired with the latest 120mm DM73 ammunition, enabling a range of engagement of up to 5,000m and offering a notable improvement over the Challenger 2's rifled gun, marking a shift to smoothbore technology more compatible with modern NATO ammunition standards. The L55A1 is an extended version of the earlier L44 gun, featuring a barrel length of 55 calibres (approximately 6.6m). This longer barrel enhances the muzzle velocity of projectiles, contributing to greater accuracy and penetrating power.

Challenger 2 Main Battle Tank Specification	
Model	Challenger 2
Manufacturer	Alvis plc, Vickers plc, BAe Systems Land & Armaments
Country	United Kingdom
Year	1998–present
Engine	Perkins CV12-6A V12 diesel 26.1L (1,590 cu in) 1,200bhp
Fuel	Diesel
Protection	Chobham/Dorchester Level 2 Classified
Top Speed	37mph (59.5km/h)
Range	340 miles (547km)
Crew Capacity	4 (commander, gunner, loader, driver)
Length	44.3ft (13.5m) (Gun forward)
Width	11.6ft (3.5m)
Height	8.2ft (2.49m)
Armament	L30A1 120mm rifled gun	7.62mm L94A1 chain gun	7.62mm L37A2 loader-operator hatch machine gun
Weight	70.5 tons (64 tonnes)
Service Branch	British Army (Royal Tank Regiment & Royal Engineers)

A Challenger 2 Main Battle Tank of A Squadron, The Queen's Royal Lancers (QRL) patrolling outside Basra in southern Iraq during Operation *Telic 4*. The driver's head can just be seen below the turret. The armour pack at the front is also visible in this image. (UK MoD)

A Challenger 2 of the Royal Scots Dragoon Guards charges across the Iraqi desert on the outskirts of Basra. This tank appears to have enhanced armour fitted to both sides of the turret as well as metal grids used to stop rocket-propelled grenades. (DPL)

Above: A Challenger 2 Main Battle Tank with new armour fitted to the hull to protect the tracks and main body of the tank as well as to the turret to safeguard the crew. The tank has electronic countermeasures fitted behind the commander. (UK MoD)

Right: The Challenger's defining feature is its 120mm main gun. The tank is capable of firing on the move at speed and identifying new targets at the same time. Smoke canisters can be seen on the right-hand side of the main gun. (UK MoD)

During war games in Oman prior to Operation *Telic*, mechanics identified that sand filters would be needed in Iraq. The exercise, *Said Sareea*, proved vital for the armoured units that deployed to Iraq, as it identified problems when operating in the desert and fixed them before 2003. (UK MoD)

CRARRV – Challenger Armoured Repair and Recovery Vehicle.

The Challenger Armoured Repair and Recovery Vehicle (CRARRV) is based on the chassis of the Challenger 1 and provided repair and recovery capability to the frontline Challenger 2 Main Battle Tanks in 2003 during Operation *Telic*. In total, 80 such vehicles had been delivered to the British Army between 1988 and 1993. The size and performance of the CRARRV are similar to a Challenger 1 tank, but instead of armament, it is fitted with a series of tools and equipment to support their role. A main winch can haul 98 tonnes, while an Atlas crane is capable of lifting 12,300lb and can lift a Challenger power pack.

Above: The additional armour package fitted to the CRARRV can clearly be seen in this image. These huge repair vehicles were manned by mechanics from the Royal Electrical and Mechanical Engineers. (UK MoD)

Left: A British Army CRARRV pictured operating in the desert. This huge vehicle was deployed to support the Warrior and the Challenger. (UK MoD)

The CRARRV is also fitted with a dozer blade to clear obstacles, while on board, the crew has access to cutting tools and a portable welder. In order to improve flexibility and supplement the transportation of power packs around the battlefield, the British Army procured a quantity of dedicated CRARRV High Mobility Trailers (HMTs). Each HMT enables a CRARRV to transport a single Challenger power pack or two Warrior AFV power packs. The reduction in the number of Challenger tanks will see a corresponding drop in the operational size of the CRARRV force.

Challenger CRARRV Specification	
Model	Challenger Armoured Repair and Recovery (CRARRV)
Manufacturer	Vickers Defence
Country	United Kingdom – also sold to Oman
Year	1988–93
Engine	Perkins-Condor CV12-5C/6C 1200bhp
Fuel	Diesel
Protection	Rolled homogeneous armour and applique
Top Speed	36mph (57km/h)
Range	310 miles (498km)
Crew Capacity	4 (commander, gunner, loader, driver)
Length	31.5ft (9.61m) (hull only); main gun removed
Width	11.9ft (3.62m)
Height	10.3ft (3.13m)
Armament	Main gun removed
Weight	67.4 Tons (61.2 tonnes)
Service Branch	British Army	Royal Electrical and Mechanical Engineers

The Scimitar is an armoured tracked reconnaissance vehicle that was deployed to Iraq in 2003 to provide surveillance and flank protection for forces moving into Basra. Its 30mm Rarden cannon allowed it to engage enemy forces and provide firepower in support of infantry. In a tragic incident during the early weeks of the invasion, one soldier was killed and five wounded when a USAF aircraft attacked two Scimitars. (DPL)

Scimitars of C Squadron, Queen's Dragoon Guards, were deployed at the Battle of Al-Faw in the opening days of the 2003 invasion of Iraq in support of the Royal Marines. They advanced ahead of the main force and engaged Iraqi forces that posed a threat to 3 Commando Brigade. These light, fast reconnaissance vehicles were deployed to monitor the flanks of the advance into Basra. (UK MoD)

The introduction of the Combat Vehicle Reconnaissance tracked (CVRT), also known as the FV107, in the 1970s delivered a new family of armoured tracked vehicles to the British Army. The driver sat to the left of the 30mm Rarden cannon and the three-man crew often strapped their combat equipment to the outside of the vehicle as space inside was limited. (DPL)

Scimitars of the Queen's Dragoon Guards played a pivotal role working closely with the Brigade Reconnaissance Force of the Royal Marines. The commander of 3 Commando Brigade, Brigadier Jim Dutton, presented the Queen's Dragoon Guards with a Commando dagger in recognition of their work.

The Scimitar was one of the most reliable and combat-capable armoured reconnaissance vehicles used by the British Army. In service, each regiment originally had a close reconnaissance squadron of five troops, each operating eight FV107 Scimitars. Its success in Iraq saw the platform deployed to Afghanistan, where it was fitted with metal guards to protect against rocket attack. However, in 2023, all Scimitars were withdrawn from service. (DPL)

The Stormer, a development of the CVRT family, included the Shielder mine-laying system, which was deployed on Operation *Telic* in 2003. It had a flat bed, which was used to mount the mine system. (UK MoD)

FV107 Scimitar – Combat Vehicle Reconnaissance (CVRT)

Scimitars of C Squadron, Queen's Dragoon Guards, were deployed at the Battle of Al-Faw in the opening days of the invasion of Iraq in 2003. These light fast reconnaissance vehicles were deployed to monitor the flanks of the advance to Basra, with C Squadron directed to support the Royal Marines amphibious landing at the Al-Faw peninsula. Fears that the area was heavily mined, however, forced the Scimitars to abandon a seaborne landing and instead cross into Iraq by land. The squadron then joined up with 3 Commando Brigade and gave vital close support to the force as it advanced across Al-Faw and towards Basra. The Scimitars worked closely with the Brigade Reconnaissance Force, operating ahead of the main brigade, and destroyed numerous Iraqi positions that threatened the marines.

The commander of 3 Commando Brigade, Brigadier Jim Dutton, presented the Queen's Dragoon Guards with a Commando dagger in recognition of their pivotal role. In one tragic incident during the early weeks of the operation, two Scimitars from the Blues and Royals were attacked by US A-10 Thunderbolt aircraft. The Blues and Royals were serving as an armoured reconnaissance element for 16 Air Assault Brigade in southern Iraq and four vehicles from D Squadron, two FV107 Scimitars and two FV103 Spartans, were moving north of the main force to patrol the forward edge of the battle area. The area of the patrol had been declared as a no-engagement zone to the allied forces and the vehicles had been given the agreed Coalition markings to avoid 'blue on blue' attacks. These included orange panels on the vehicles, plus thermal reflectors and the flying of the Union flag.

The introduction of the Combat Vehicle Reconnaissance Tracked (CVRT), also known as the FV107, in the 1970s delivered a new family of armoured tracked vehicles to the British Army. The Scimitar armed reconnaissance platform, sometimes called a light tank, was fast and was often used to ferry soldiers on its chassis over short distances.

The Scimitar family of CVRTs included nine variants. The FV104 Samaritan was the armoured ambulance used by the British Army in the Iraq War. Manned by soldiers from the Royal Army Medical Corps, it provided protected mobility to evacuate wounded soldiers from the battlefield. Inside, the vehicle space was tight, with four stretchers and a narrow passageway for medics to treat patients.

The FV105 Sultan was the command vehicle based on the CVR(T) platform. It has a higher roof than the Armoured Personnel Carrier (Spartan) variants, providing a more comfortable 'office space' inside. The Sultan entered service in 1978 and served with the 7th Armoured Brigade.

The FV102 Striker was the anti-tank guided missile carrier version of the CVR(T) and was developed to launch the Swingfire missile. Delivered in 1975, the first production vehicles were employed by the Royal Artillery's anti-tank guided missile batteries. On 24 March 2003, during Operation *Telic*, a Striker destroyed an Iraqi T-55 tank with an anti-tank missile. With the replacement of the Swingfire missile by the Javelin in mid-2005, the FV102 Striker was withdrawn from British Army service.

The FV106 Samson was the recovery platform of the CVR(T) family of vehicles, conceived in the early 1970s with the final design entering production in 1978. As well as being able to recover CVRTs, it could also recover other light tracked vehicles, such as the older 430 series. Its hull is an all-welded aluminium construction.

The FV103 Spartan was the tracked armoured personnel carrier of the CVR(T) family and was deployed by the British military in 1978. With the exception of the Striker's missile launcher, the Spartan shares a similar appearance with the FV102 Striker. It can accommodate seven people, comprising two crew members and five passengers or three crew members and four passengers.

Finally, the Stormer development of the CVR(T) family included the Shielder mine-laying system deployed in Operation *Telic* in 2003.

The Scimitar reconnaissance vehicles were often deployed to provide support to infantry troops as unofficial battlefield taxis, with soldiers sitting on the chassis for short distances. Once the main phase of the war was over, the role of the Scimitar and many of the CVRTs was reduced. Although it could be fitted with protection against rocket-propelled grenades, it was not suited to heavy plate armour.

Scimitar – Combat Vehicle Reconnaissance (Tracked) Specification	
Model	Scimitar, Scorpion, Striker, Spartan, Samaritan, Sultan, Samson, Shielder
Manufacturer	Alvis | BAe Systems
Country	United Kingdom
Year	1970s–2022
Engine	Jaguar J60 4.2L petrol | later replaced with Cummins BTA 5.9L diesel
Fuel	Petrol | Diesel
Protection	Aluminium alloy armour
Top Speed	50mph (80.5km/h)
Range	300 miles (450km)
Crew Capacity	3–7 depending on variant
Length	16.07ft (4.9m)
Width	7.2ft (2.2m)
Height	6.88ft (2.1m)
Armament	30mm Raden cannon on Scimitar | 76mm gun on Scorpion
Weight	8.5 tons (7.8 tonnes)
Service Branch	British Army | Light Recce

FV510 – Warrior (Tracked Armoured Personnel Vehicle)

The Warrior FV510 armoured personnel carrier was among the first vehicles to be shipped to the Middle East in 2003. Able to carry seven soldiers, the 24-tonne vehicle was fitted with a 30mm Raden cannon and had the ability to reach speeds of 45mph. Inside, the forward right compartment houses the driver's area, while the gunner and commander are both seated in the turret. The seven infantry soldiers in the rear sit facing each other; they exit and access through the single electric ram-powered back door. The Warrior was also equipped with Clansman radios, replaced in the 1990s by Bowman. Also, each Warrior comes with passive image intensifier night-vision sights.

The Warrior's forward left compartment houses a Perkins-Rolls-Royce V8 Condor engine, transmitting to front sprockets through a four-speed automatic gearbox. Top speed was 46mph and the vehicle was regarded as capable of going anywhere. To handle the most difficult terrain, it had a set of six roadwheels and three return rollers per side plus rear idler. The forward hatch above the transmission and power pack can be removed and replaced in less than one hour by two men.

Fitted with enhanced applique armour, the Warrior was deployed in Basra by the 1st Battalion Irish Guards. Each vehicle was fitted on the outside with the invasion markings and Combat Identification system to avoid 'blue on blue' attacks by Coalition aircraft.

Warrior variants deployed in Operation *Telic* included the V510 Infantry Fighting Vehicle, the FV511 ICV: Infantry Command Vehicle and the FV512 MCV: Mechanised Combat Repair Vehicle, which was used by detachments of specialist mechanics in support of Warriors and Challengers. In addition, the FV514 Mechanised Artillery Observation Vehicle (MAOV) was deployed for the Royal Artillery as an Artillery Observation Post Vehicle (OPV).

Warrior Armoured Fighting Vehicles provide static protection during an incident on the outskirts of Basra City. The Warrior was fitted with Chobham armour and metal 'anti-rocket' grilles. The driver's position is to the left at the front and his armoured cover is closed to protect him from snipers. (UK MoD)

A Warrior of the Welsh Guards battle group at speed on a desert road north of Basra. The vehicle still bears the Stabilisation Force (SFOR) identification that was carried by NATO vehicles in the Balkans. At the front, a vertical post was fitted to cut any wires placed across the road with the intention of injuring crew or igniting an IED. (DPL)

The UK exported the successful modified version called Desert Warrior to Kuwait. This included a Delco turret (used on the LAV-25) with the stabilised M242 Bushmaster 25mm chain gun, plus a 7.62mm coaxial MG and two Hughes TOW Anti-Tank Guided Missiles (ATGMs). One requirement was a top speed sufficient to keep up with what was then the new main battle tank, the Challenger, as the FV430 armoured personnel carrier and variants could not match its top speed.

The first pre-production Warrior models were delivered in November 1984. After trials, it was accepted into service by the British Army. In total, 789 FV510 and variants were produced for the British Army, and 254 more of the Desert Warrior for the Kuwaiti Army. Since then, the Warrior has been upgraded numerous times and, after the Iraq War, went on to serve in Afghanistan. Plans to make further enhancements to the Warrior force, called the Warrior Capability Sustainment Programme, were suspended in 2020, and the Warrior will eventually be replaced by the Ajax armoured platform.

Warrior – Tracked Armoured Vehicle Specification	
Model	Warrior FV510
Manufacturer	GKN Sankey/GKN Defence
Country	UK
Year	1986–present
Engine	Perkins V8 Condor Diesel 550hp
Diesel	Diesel
Protection	Aluminium and appliqué
Top Speed	46mph (75km/h)
Range	410 miles (660km)
Crew Capacity	3 plus maximum of 10 passengers
Length	20.8ft (6.3m)
Width	9.11ft (3.93m)
Height	9.2ft (2.8m)
Armament	30mm L21A1 Rarden cannon
Weight	27.9 tons (25.4 tonnes)
Service Branch	British Army

Warriors serving with the 1st Battalion, The Princess of Wales's Royal Regiment depart their base at Basra Air Base in May 2006. Warriors always operated in pairs to provide mutual protection in case of an attack or breakdown. The enhanced Chobham armour packs can clearly be seen on these Warriors, plus the metal grilles fitted to detonate any rockets before they get a chance to hit the vehicle. (UK MoD)

Warriors serving in support of the Welsh Guards mount a patrol on the perimeter of the Coalition airfield at Basra. Both drivers' hatches are slightly open, no doubt due to the intense heat. The Warrior was mainly used as a battlefield taxi and could carry ten troops in the rear. (DPL)

A Warrior Infantry Fighting Vehicle of the Black Watch Battle Group bears the scars of several rocket-propelled grenade attacks on its Chobham armour while on operations in Iraq during October 2004. The enhanced armour absorbed the kinetic energy of the RPGs' explosive blasts. (UK MoD)

A FV512 recovery Warrior deployed in support of both Warrior and Challenger vehicles. The 512 was manned by teams from the Royal Mechanical and Electrical Engineers and carried out engine and gearbox changes in the desert. (UK MoD)

Infantry soldiers dismounted from the Warrior via a rear door. This image shows soldiers deploying from a Warrior during a training exercise at Copehill Down exercise area on Salisbury Plain. (DPL)

The first production vehicle was handed over to the British Army in May 1987 to 1st Battalion Grenadier Guards and since its introduction it has undergone numerous upgrades. The vehicle, which is due to be replaced by the Ajax, was intended to undergo a major sustainment programme, but this was cancelled by the UK in 2021. (UK MoD)

FV430 Bulldog Armoured Personnel Carrier

The British Army's Bulldog is the upgraded variant of the FV432 armoured personnel carrier, first introduced in the 1970s. It is officially listed as the FV430 and was first deployed to southern Iraq in 2007 with the 1st Battalion, Royal Green Jackets, in Operation *Telic 9*. The vehicle was fitted with appliqué reactive armour capable of defeating hollow-charge warheads such as rocket-propelled grenades (RPGs), which were widely used by insurgents in Iraq. A new engine and steering system provided increased mobility and manoeuvrability. Other features include air conditioning, and a gun station, mounted on top of the Bulldog, fitted with a 7.62mm machine gun that can be controlled from inside the vehicle. The Bulldog has two crew members and is capable of carrying up to ten passengers with a maximum range of 480km. Bulldog variants are in service with the infantry as command vehicles, 81mm mortar carriers, ambulances and recovery vehicles. The vehicles can also be converted for use in water and also have strong cross-country performance.

The FV432, which the Bulldog replaced, entered service in the 1970s and became the workhorse of armoured infantry operations. It was due to be phased out as the Warrior entered service, but its robust capability ensured its survival. More than 3,000 vehicles were produced by GKN Sankey and delivered to the British Army. This powerful APC was fitted with a Rolls-Royce water-cooled petrol engine, giving the vehicle a top speed of 32mph on the road. It was an all-steel construction; the chassis was a conventional tracked design with the engine at the front and the driving position to the right. Directly behind the driver's position is the vehicle commander's hatch. There is a large round opening in the passenger compartment roof, which has a split hatch, and a side-hinged door in the rear for loading and unloading. In common with many designs of its era, there are no firing ports for the troops carried; British Army doctrine has always been to dismount from vehicles to fight, whereas Soviet/Russian infantry fighting

British troops drive the first upgraded FV430 Mk3 Bulldog vehicles into Basra. The vehicle featured a bulletproof shield for the gunner, although his back was exposed to snipers. (UK MoD)

vehicles largely incorporate weapon ports. The Bulldog's passenger compartment has five seats on each side; these fold up to provide a flat cargo space. With a capacity for ten and often more, it would deliver them as close to a battle as possible. Its only armament was a single 7.62mm GPMG (General Purpose Machine Gun) plus smoke dischargers fitted to the front armour, while the Bulldog can also be fitted with a heavy-calibre machine gun.

Bulldog – FV 432 Tracked Armoured Personnel Carrier Specification	
Model	FV432 Bulldog
Manufacturer	BAE Systems Land Systems
Country	United Kingdom
Year	2006–present (upgraded variant)
Engine	Cummins Engine Company B-series six-cylinder turbocharged diesel engine developing 250hp
Fuel	Multi-fuel
Protection	Enhanced reactive armour
Top Speed	32mph (52km/h)
Range	360 miles (580km)
Crew Capacity	2 plus 10 passengers
Length	17.3ft (5.25m) (hull only); main gun removed
Width	8.4ft (2.55m)
Height	7.6ft (2.28m)
Armament	7.62mm General Purpose Machine Gun
Weight	16.5 tons (15 tonnes)
Service Branch	British Army

The armoured panels on the side of the Bulldog gave much greater protection to the crew and personnel inside. The Bulldog was also fitted with electronic countermeasures to identify IED threats. (UK MoD)

A Bulldog moves down a street in Basra city in support of infantry soldiers on foot patrol. While the Bulldog was a major improvement, the driver and commander remained exposed. (UK MoD)

Right: The armoured protection covering the Bulldog's tracks can clearly be seen in this image. The angled armour on the side was designed to deflect rocket-propelled grenades (RPGs). Its enhanced protection was welcomed by soldiers. (UK MoD)

Below: Three Bulldog armoured vehicles of A Company, 2 Rifles, which were serving with 19 Light Brigade in Basra, pause on the outskirts of the city. The metal grilles to stop RPGs can be seen at the rear of the Bulldog. (UK MoD)

FV180 Combat Engineer Tractor

The FV180 Combat Engineer Tractor or CET was a specialist armoured vehicle deployed by the British Army to Iraq in 2003. This small, highly manoeuvrable tracked armoured vehicle had an amphibious capability and was used by the Royal Engineers. It was employed on ground preparation for bridge construction and towing activities in the front line of battle, such as digging vehicle fighting pits, constructing earthen barriers, repairing roads, recovering disabled vehicles from water and other obstacles, preparing riverbanks for vehicle crossings and clearing obstacles.

The two crew sit in tandem positions on the left-hand side of the vehicle, each with a set of driving controls facing opposite directions. A large earth-moving bucket is fitted at the rear of the vehicle and a rocket-propelled anchor on a 100m hawser attached to an eight-tonne winch can be fitted to the front. When operated from the rear seat, the bucket is used for earth moving; clearing obstacles, making paths or digging tank or gun pits and anti-tank ditches. When operated from the front-facing seat, it can be driven on the road, and the anchor can be used to pull the CET up steep obstacles such as riverbanks. The winch rope can be deployed to the front or the rear of the vehicle with a maximum pull of eight tonnes in both configurations.

The vehicle is protected for operations in a nuclear, biological and chemical (NBC) environment. The NBC air system is also used to inflate the buoyancy aids required to trim the vehicle when swimming. The amphibious propulsion is provided by two Dowty water impellers, one mounted on each side of the vehicle and controlled by the commander in the rear seat facing forwards. The water jets are used to steer the vehicle when swimming, additional to the use of movable cowls directing the flow of water. When not in use, the propulsion unit water intakes are closed off with armoured

The FV180 Combat Engineer Tractor or CET was a specialist armoured vehicle deployed by the British Army to Iraq in 2003. It was fast, reliable and had an amphibious capability. (UK MoD).

covers to prevent damage during digging operations. Flotation aids are required to trim the vehicle for swimming and a 'wash board' is fitted to the front of the vehicle to prevent the bow wave flooding the crew compartment when entering water. Maximum speed in water is 8.5 knots; the vehicle will wade in 1.8m of water, but requires preparation for operating in deeper water than this as it achieves buoyancy. The FV180 can tow a 'Giant Viper' anti-mine system, and can be airlifted to forward operations by a C-130 Hercules aircraft.

Combat Engineer Tractor (CET) Armoured Specification	
Model	FV180 CET
Manufacturer	Royal Ordnance (UK)
Country	United Kingdom
Year	1977–87 (Total of 141 built)
Engine	Rolls-Royce C6TFR
Fuel	Diesel
Protection	Honeycombed, twin-skin aluminium alloy armour
Top Speed	35mph (56km/h)
Range	300 miles (480km)
Crew Capacity	2
Length	24.9ft (7.54m)
Width	9.8ft (2.94m)
Height	8.9ft (2.67m)
Armament	Crew personal weapons only
Weight	19.2 tons (17.5 tonnes)
Service Branch	British Army – Royal Engineers

TPz Fuchs – Nuclear, Biological and Chemical Reconnaissance Vehicle

Fuchs vehicles from the Joint Nuclear, Biological and Chemical warfare unit were deployed in Operation *Telic*, the invasion of Iraq in 2003, and provided support to US forces. Concern that Saddam Hussein would use chemical weapons against Coalition forces during Operation *Iraqi Freedom* resulted in the US and British Army seeking additional protection, and the German-produced Fuchs vehicles were deployed to support ground reconnaissance of chemical threats and electronic warfare. The six-wheeled armoured vehicles offered total protection in a nuclear, biological and chemical environment and was vital to UK forces amid claims that Iraqi forces would deploy Sarin gas.

Developed by Germany's Daimler-Benz and Rheinmetall MAN Military Vehicles (RMMV), the vehicle had been ordered for the Bundeswehr (West German Army). As well as its off-road capability, the Fuchs is fully amphibious with a maximum water speed of ten knots. Additional to its primary NBC role, the platform can also be used for electronic warfare, medical evacuation, transportation of a mortar team and for bomb-disposal tasks. The Fuchs's hull is made entirely of welded armoured steel. Inside, the vehicle commander is seated to the right of the driver; the commander and driver each have their own door. There are metal shutters on the windscreen and door windows that can be closed to seal the vehicle. Periscopes installed in the vehicle's roof, forward of the driver's hatch, provide a view outside the cabin. There is a circular roof hatch for the commander.

NBC-protected Fuchs vehicles from the Joint Nuclear, Biological and Chemical warfare unit were deployed in Operation *Telic*, the invasion of Iraq, in 2003. They operated in areas ahead of the force to check the air samples in case of gas attack by Iraq. (UK MoD)

Prior to deployment, numerous training exercises took place. The six-wheeled armoured vehicles offered total protection in a nuclear, biological and chemical environment and were vital to UK forces amid claims that Iraqi forces would deploy Sarin gas. (UK MoD)

During Operation *Granby*, the UK's part of Operation *Desert Storm*, the Fuchs vehicles played a key role in testing and evaluating air samples. After the war, the UK formed a joint NBC squadron of RAF and Royal Tank Regiment personnel to provide specialist support across the British armed forces. In 2020, the UK Ministry of Defence announced a £16 million contract to upgrade and sustain the British Army's critical fleet of specialist NBC-capable vehicles as well as those used on surveillance and reconnaissance. The Fuchs vehicles in question are six-wheeled, all-wheel drive armoured vehicles that have been customised to a protected configuration to carry out chemical, radiological and nuclear survey and reconnaissance missions. The vehicles are equipped with automatic systems and sensors for detecting nuclear radiation as well as Chemical, Biological, Radiological and Nuclear (CBRN) agents and other toxic substances. The nine-strong fleet of vehicles is complemented by a training simulator, also to be updated and sustained under the contract, which ensures that the specialist operators within the Warminster-based Falcon Squadron can undergo regular training on site.

FUCHS – Armoured NBC Reconnaissance Vehicle Specification	
Model	TPz Fuchs – Nuclear, Chemical Biological Reconnaissance (9 in service)
Manufacturer	Daimler-Benz
Country	Germany
Year	1979–present
Engine	Mercedes-Benz Model OM402A V8 liquid-cooled diesel engine
Diesel	Diesel
Protection	Steel armour
Top Speed	65mph (105km/h)
Range	500 miles (800km/h)
Crew Capacity	2 plus ten passengers
Length	22ft (6.8m)
Width	9.8ft (2.98m)
Height	8.2ft (2.5m)
Armament	MG3 Rheinmetall machine gun
Weight	18.7 tons (17 tonnes) (empty) 25.3 tons (23 tonnes) (combat loaded)
Service Branch	British Army Falcon Squadron

Ridgback and Mastiff (MRAPs)

The increase in roadside bombs during the early years of the counter-insurgency forced a review of vehicle protection in Basra where the British Army's main force had operated since 2003. In 2005, the UK Ministry of Defence (MOD) placed an Urgent Operational Requirement (UOR) for 157 Cougar all-terrain wheeled vehicles with the US-based company Force Protection (later acquired by General Dynamics in 2011).

The Cougar four-wheeled variant was renamed Ridgback and the six-wheeled variant was renamed the Mastiff. The vehicles were heavily armoured and some were equipped with the Enforcer remote weapon system. They also included an NBC overpressure system to protect against any biological attack. The electrically powered winch had the capacity to haul 9,000lb, while run-flat tyres gave the vehicle added capability. The first batch of five Ridgback was delivered to RAF Brize Norton on 14 August 2008 and quickly flown to Iraq.

In contrast to the original Cougar vehicle, the British Mastiff included vertical armour plates covering the weapon firing ports and large visibility blocks. Mastiff has a maximum speed of 55mph, and was armed with a 7.62mm general purpose machine gun, 12.7mm heavy machine gun or 40mm automatic grenade launcher. Inside, it had Bowman radios and electronic countermeasures. Both the Ridgback and Mastiff were praised by British soldiers as game-changers. As attacks soared in Basra, soldiers preferred to travel in the vehicles, which were dubbed 'safe boxes' due to their heavy armour and robust reputation. The vehicles remain in service with the British Army and were also deployed to Afghanistan.

MRAP (Mine Resistant Ambush Protected) Specification		
Model	Ridgback (4x4)	Mastiff (6x6 Variant)
Manufacturer	General Dynamics Land Systems (Formerly Force Protection Inc)	
Country	United States/United Kingdom	
Year	2002 (produced); in service 2004–present	
Engine	Caterpillar C7 diesel 330hp	
Fuel	Diesel	
Protection	Classified	
Top Speed	65mph (105km/h)	
Range	600 miles (966km)	
Crew Capacity	2 plus 4 passengers	2 plus 8 passengers
Length	19.41ft (5.91m)	23.25ft (7.08m)
Width	9ft (2.74m)	
Height	8.67ft (2.64m)	
Main Armament	M240 7.62mm machine gun or M2 .50 calibre machine gun or optional remote weapons system	
Weight	15.9 tons (14.5 tonnes)	25 tons (22.7 tonnes)
Service Branch	British Army	

Two six-wheeled Mastiff Protected Patrol Vehicles deployed with bomb disposal team of 11 Explosive Ordnance Detachment (EOD) Regiment, Royal Logistic Corps. The Mastiffs were fitted with Bowman radios and electronic countermeasures. As well as their heavy armour protection, the Mastiffs were fitted with metal grilles to counter (RPGs). (UK MoD)

A Mastiff, driven by soldiers of the Royal Logistics Corps, is loaded into an RAF C-17. The aircraft could carry two Mastiff vehicles. These huge wheeled vehicles were one of the safest types used by British forces in Basra. Vehicles were attacked, but in the main survived intact. A vertical bar on the right and left of the front window was fitted to cut any wires across the road that could be used to ignite an IED. (UK MoD)

A convoy of three British Mastiff armoured vehicles on patrol in southern Iraq during Operation *Charge of the Knights-14* in Basra city. The mission was undertaken by the UK Military Transition Team (MITT) Group attached to 50 Brigade of the New Iraqi Army (NIA). Its aim was to support the NIA as the deadline for UK forces' withdrawal approached. (UK MoD)

Mastiff and Ridgback vehicles were ordered as part of an Urgent Operational Requirement (UOR) and underwent extensive testing before being deployed to Basra in southern Iraq. (David Pimbblet/DPL)

Left: A Ridgback Protected Vehicle. The MoD has ordered 157 Cougar 4x4s from the United States, as part of an UOR. Having arrived in the UK, the vehicles were upgraded with integrated additional protection, weapons, communications systems and specialist electronic countermeasures. They then joined their Mastiff big brothers on operations. (UK MoD)

Below: A Ridgback ambulance variant. The high level of attacks in Iraq had forced the Coalition nations to focus on 'protected mobility' to improve safety for soldiers. The Ridgback often operated alongside the Mastiff on operations. (UK MoD)

Chapter 3
Iraq

Iraqi Army

After the 1991 Gulf War the UN Security Council Passed Resolution 687, which directed Iraq to destroy all its weapons of mass destruction (WMDs) – a term used to describe nuclear, biological and chemical weapons, as well as long-range ballistic missiles. The Iraqi leader replied that his country had no such weapons. US President George W. Bush insisted that Saddam was continuing to produce WMDs, going on to state that Iraq was part of an 'axis of evil' in common with North Korea and Iran. In October 2002, the US Congress authorised the use of military force against Iraq, which sparked controversy in Europe and the Middle East.

In Baghdad and cities across Iraq, United Nations weapons inspectors had been sent to locate these WMDs following Washington's ongoing claims that Saddam and his leadership was harbouring 'biological and chemical weapons' and was ready to use them against the West. The weapons inspectors found nothing to substantiate the President's claims, but asked for more time to complete their work. Washington said no. In the weeks prior to the invasion, a propaganda war filled the television screens; Iraqi Minister of Information Muhammad Saeed al-Sahhaf, nicknamed Comical Ali or Baghdad Bob, made frequent screen appearances in which he announced bombastically that the Americans would be sent home in body bags if they invaded Iraq. He also claimed that Iraq had built a special tank and would field more than 600 of these vehicles to destroy the Coalition before it could reach the capital.

There is no question that for thousands of Iraqis, life under Saddam had been hard, with secret police often arresting and detaining people suspected of not showing 100 per cent support for the regime. Under Saddam Hussein, Iraqi's Sunni ruling class dominated the country's governing body and National Assembly. Likewise, the Ba'ath Party managed a brutal reign over the marginalised Shia community. Considerable numbers of oppressed Iraqis wanted to see the end of Saddam, with Bush seeing his own mission as doing just that in order to give the people their freedom, However, although the people did indeed crave liberty, they did not particularly want the Americans on the streets of Iraq.

While the Iraqi Army had suffered heavily against Coalition forces in the 1991 Gulf War, it still had a considerable number of tanks, which included T-72 and T-62s, the vanguard of Iraq's frontline forces. As President George W. Bush gave his forces approval to strike Iraq alongside those of the UK, it was unclear if Saddam would use chemical weapons. Coalition tank crews crossed into Iraq with their cupolas closed down, while infantry troops either wore chemical protection suits or had them ready for use. The fear was that the Iraqi artillery would fire shells loaded with chemical or biological gas, which it was eventually confirmed they did not have. Many of Saddam's tank units were not operationally effective and sat in a defensive line across the desert south of Baghdad. Even so, Iraq's hard-line Republican Guard and Fedayeen forces put up a defence of Baghdad and Basra and fought fierce battles with US and UK forces. In one incident, a US Abrams tank was hit by a rocket-propelled grenade (RPG) on the outskirts of Baghdad, setting the engine on fire and forcing the crew to abandon the vehicle.

T-72 Main Battle Tank

Iraq's armoured force was spearheaded by the T-72s of the elite Medina Division. Among these Soviet-built main battle tanks (MBTs) was an Iraqi-built version known as the Asad Babil or Lion of Babylon.

These tanks were assembled at a factory in Taji, north of the capital, which was established in the 1980s in an attempt by the regime to develop an indigenous tank. This strategy had come about after Western governments imposed an embargo on Iraq in order to force a negotiated end to the Iran–Iraq war.

Developed in the 1960s, the T-72 was known for its robust design, powerful 125mm smoothbore gun and composite armour. Iraq acquired a substantial number of T-72 tanks from the Soviet Union in the 1970s and 1980s, making it a central component of Saddam Hussein's military capability. Compared to MBTs used in the West, the T-72 has a much smaller profile and is lighter at 41 tonnes. Even so, it was highly capable and could traverse rivers entirely submerged, using an installed snorkel. For such operations, each member of the crew is equipped with a basic chest-pack rebreather in case of emergency. If the engine stops underwater, the T-72's engine compartment floods from pressure and the engine has to be started again in six seconds. The limited space made escape challenging, so 'submerge' exercises were not popular. The tank is also NBC capable and to reduce penetrating radiation from neutron bomb explosions, boron-compound synthetic cloth is used to line the inside of the hull and turret, while an air-filter system provides clean air to the crew. A small amount of overpressure keeps contaminants from entering through joints and bearings. The main gun's autoloader facilitates more effective forced smoke removal than conventional manually loaded tank guns, so that, in theory, NBC isolation of the fighting compartment can be maintained forever.

Similar to other Soviet-era tanks, the T-72's design sacrifices interior space to achieve a very thin profile, even going so far as to swap out the fourth crew member for a motorised loader. The interior space of the T-72 is limited by the low height of the tank; therefore, the Soviet Army set a maximum

A T-72 Asad Babil (Lion of Babylon) abandoned almost intact on the US forces' main resupply route in Iraq, 2003. Some of the tank's main features, like the reinforced glacis plate and the bracket for the electro-optical countermeasures pod (removed) are clearly visible in this photograph. (US DoD)

height limit of 5ft 4in for crew members. Even by the restricted standards of battle tanks, the original T-72 design has incredibly small periscope viewports and, when the hatch is closed, the driver's field of vision is very limited. Rather than the more user-friendly steering wheel or steering yoke found in many contemporary western tanks, the steering system is a dual-tiller arrangement.

The 1973-built turret of the first T-72 is constructed entirely of traditional cast high-hardness steel (HHS) armour, and every subsequent iteration saw improvements to the tank's armour protection. The front plate measures 3.1in, with the maximum thickness is estimated to be 11in. For all of this, Iraq's T-72s didn't fare well against the new and improved Western armour during the 2003 invasion. Having already experienced the effectiveness of the M1A1 Abrams during the 1991 Gulf War, the Iraqi armour stood little chance when the US returned to the region. The Republican Guard's T-72s, the majority of which were from the Medina Division, were stationed across Baghdad during the 2003 invasion of Iraq, their role being to defend the seat of the Ba'ath regime.

When US tanks confronted Iraqi T-72s in close quarters in March 2003, they destroyed seven without suffering any casualties. These encounters revealed Iraqi gunners' subpar shooting, partly due to a lack of sophisticated night-vision and rangefinder equipment. It is unclear whether the tanks' technological capabilities or crew training had improved between the Persian Gulf War and this conflict; there was already a noticeable gulf in quality between Iraq's Soviet-derived equipment and that of the Coalition of the time. Nevertheless, when Iraqi armoured forces tried to strike their American opponents close to Baghdad airport in 2003, one Bradley was severely damaged by a 125mm round fired by an Asad Babil tank. Following waves of American armoured advances on the capital of Iraq, the final functioning T-72s that had been obtained from the USSR and Poland were destroyed, or their operators abandoned them after the fall of Baghdad; several left the battlefield without firing a single shot. Afterwards, US Army disposal troops dismantled the abandoned tanks or sent them to the US for target practice.

T-72 – Main Battle Tank (Numerous Variants) Specification	
Model	T-72 MBT
Manufacturer	Soviet Union (Russia) Kartsev-Venediktov and Lion of Babylon project
Country	Iraq
Year	1989–90
Engine	V-12 diesel 780hp
Fuel	Diesel
Protection	Steel and composite armour with ERA
Top Speed	40mph (64km/h)
Range	290 miles; extended to 430 with external fuel drums (466–692km)
Crew Capacity	3 (commander, driver, gunner)
Length	31.11ft (9.73m) (gun forward)
Width	12.9ft (3.89m)
Height	8.11ft (2.73m)
Armament	125mm 2A46M/2A46M-5 smoothbore gun; support from 12.7mm heavy machine gun
Weight	51.2 tons (46.5 tonnes)
Service Branch	Iraqi Army

Above: Iraqi T-72 MBT recovered by the United States and on show at the US Army's 1st Cavalry Division Museum. (US DoD)

Left: Iraqi soldiers stand guard on the streets of Baghdad in a T-72. This tank was the most threatening to the Coalition of all Saddam's armour. (Iraqi Govt)

US Army engineers use an M88 platform to recover the remains of a T-72 in August 2003 during a massive 'clear-up' of the Iraqi countryside. (US DoD)

T-62 Main Battle Tank

In 2003, the T-62 medium tank, known under the Soviet identification of Object 166, formed a large proportion of Iraq's armoured forces. Designed and built at Factory No 183 (Uralvagonzavod) in Nizhniy Tagil, it had been developed as a direct response to the new American M60 tank, which had been dispatched to the 3rd Armoured Division to serve with the US Army in Europe in December 1960. Officially entering service with the Soviet Army in August 1961, the T-62 was an amalgamation of several existing concepts that were well established in the USSR before the M60 was known about, but had since been taken forward from the experimental stage.

Work on a new medium tank programme had begun in 1953 and research already accumulated was used over the next few years to shape the design into what became the T-62. Many standard components, from communications to lighting, were carried over from previous tanks, and to that end, the crew of a T-62 was equipped with the same controls and observation devices as those of their counterparts manning the previous T-55s. The driver was provided with two periscopes, laid out to ensure that he could see both front corners of the hull. He could swap out one of them for a night-vision periscope,

Purchased from the Soviet Union, the Iraqi T-62 fleet had been deployed in the war against Iran and many tanks were in need of maintenance and upgrading. Even so, Iraq still had hundreds of tanks and posed a serious threat to the US-led Coalition. (DPL)

Above: An Iraqi T-62 alongside an MT-LB armoured personnel carrier. After the war, a review highlighted that, while some Iraqi tank forces fought well, the majority of armoured crew lacked tactical awareness. (DPL)

Left: Buried in the desert winds, this T-62 was clearly abandoned. The spanner in the left-hand side of the image suggested that the crew had been carrying out maintenance when the tank was hit in a Coalition attack. (US DoD)

which could also be mounted externally when driving from an open hatch. The loader had a single MK-4 rotating periscope for a relatively restricted view towards the left side of the turret. The gunner was provided with a single forward-facing periscope for general observation and to alleviate car sickness, while his main observation device was the TSh2B-41 telescopic sight.

The on-board fuel carried in a T-62 was divided between four internal Bakelite-coated steel tanks, holding 675L, and three external tanks on the fenders with a capacity of 285L, for a total capacity of 960L. Additionally, a pair of external 200L fuel drums could be mounted onto the rear of the hull for extended range.

Being a development of the T-55 series, the T-62 retained much of the low profile and substantial turret armour of its predecessor. The T-62 was the first production tank equipped with a smoothbore tank gun that could fire APFSDS rounds faster than earlier tanks, which were mounted with rifled tank guns. Due to its greater manufacturing costs and maintenance requirements by comparison with its predecessor, the T-62 did not completely replace the T-55 in export markets, even though it became the standard tank in the Soviet arsenal.

Inside the T-62, space was at a premium, with the driver's compartment located in the lower front, the fighting compartment in the middle and the engine compartment at the rear. The loader, gunner, driver and commander make up the crew of four. As well as being one of the world's first tanks to introduce a smoothbore gun, it used APFSDS ammunition as its standard armour-piercing ammunition, this concept being adopted from the T-12 towed anti-tank gun. The T-62's 115mm main gun had the factory designation of U-5TS, and the tank had the space to carry 40 'immediate use' rounds, with additional rounds kept in storage in the front of the hull, to the right of the driver, and in the rear of the fighting compartment. On operations, four rounds are kept in the turret, while the coaxial machine gun's 2,500 rounds are also stored inside.

The T-62 was less manoeuvrable than the T-55 because of its increased weight. Similar to the T-55, the T-62 is equipped with a single auxiliary oil tank on the left fender and three external diesel fuel tanks on the right fender.

T-62 – Main Battle Tank (Numerous Variants) Specification	
Model	T-62 | Main Battle Tank | variants
Manufacturer	Soviet Union (Russia) Uralvagonzavod
Country	Iraq
Year	1961–present
Engine	V12 diesel 780hp (582kw)
Fuel	Diesel
Protection	Heavy protected turret and upper side with 153mm steel
Top Speed	31mph (49km/h)
Range	280 miles (450km)
Crew Capacity	4 (commander, loader, driver, gunner)
Length	30.8ft (9.34m)
Width	10.10ft (3.30m)
Height	7.10ft (2.40m)
Armament	Main gun 115mm smoothbore U-5TS (2A20) barrel, support from 12.7mm heavy machine gun and 7.62mm PKT coaxial machine gun
Weight	40.7 tons (37 tonnes)
Service Branch	Iraqi Army

Type 69-QM

Iraqi Army units defending Nasiriyah in March 2003 were equipped with Chinese-supplied Type 69-QMs, the majority of which were pressed into service as artillery pillboxes before their destruction by AH-1 Cobra helicopters. These tanks were a key component of the ambushes launched against Charlie Company of the 1st Battalion, 2nd Marines, and the US Army 507th Maintenance Company. Two Type 69s destroyed at least four vehicles of the 507th, one of which was a large truck. Another first-hand story describes four Type 69s concealed behind buildings, hitting Charlie Company with indirect fire and taking out a few AAVs. Certain combat-ineffective Type 59/69s were positioned as obstacles or decoys.

The Type 69-QM in Iraqi service was fitted with a 100mm rifled gun as standard and a secondary coaxial machine gun, along with a 12.7mm anti-aircraft machine gun. According to Western observers, Iraq modified some Type 69s with a 105mm cannon, known as the QM1, and a 125mm gun with an autoloader, known as the QM2. The Type 69-QM was essentially a Chinese Type 69-II modified for Iraq

Above: The Type 69 was an old tank at the time of the war in 2003. Spares had been a serious issue for the Iraqi army and when the US Coalition invaded, many units equipped with the Type 69 deployed them in a static role. (US DoD)

Left: A large number of Chinese-made Type 69s was deployed by the Iraqi Army to Nasiriyah. The Type 69-QM was essentially a Chinese Type 69-II modified for Iraq by fitting it with a laser rangefinder, infrared night vision and a computerised fire-control system. (US DoD)

through the fitting of a laser rangefinder, infrared night vision and computerised fire-control system. Diesel fuel could be poured onto the exhaust when it was hot, to create a makeshift smokescreen to conceal the tank from enemy forces.

Type 69 – Main Battle Tank Specification	
Model	Type-69 QM | Main Battle Tank
Manufacturer	China – First Inner Mongolia Machinery Factory, Norinco
Country	Iraq
Year	1959–present
Engine	12150L-7 V12 diesel engine 580hp (430kW)
Fuel	Diesel
Protection	203mm
Top Speed	30mph (38km/h)
Range	270 miles (434km)
Crew Capacity	4 (commander, gunner, driver, loader)
Length	20.4ft (6.24m)
Width	10.8ft (3.3m)
Height	9ft (2.7m)
Armament	Main gun 100mm smoothbore/105mm rifled
Weight	40.4 tons (36.7 tonnes)
Service Branch	Iraqi Army

BMP-1 Armoured Personnel Carrier

When the BMP was drawn up in the late 1950s by the Soviets, it was required to be fast, have a powerful weapon and allow every squad member to shoot from inside the vehicle. The BMP needed to be able to defeat similar small, armoured vehicles as well as directly help dismounted infantry in both attack and defence. The vehicle had to shield the crew from light shell fragments at 500–800m away, as well as from 20–23mm autocannons and .50-cal machine-gun fire across the frontal arc. Firepower was provided by an anti-tank wire-guided missile (ATGM) launcher and a 73mm 2A28 Grom cannon. The gun's planned range was listed at 700 metres (770 yards), which was enough to engage enemy armoured vehicles and ground forces. The missile launcher was designed to be employed against targets at ranges between 500 and 3,300m.

The BMP-1 features an electric traverse drive with a manual back-up mechanism and a fume extractor system installed in its conical turret. There is a dead zone over the commander's hatch where the main cannon must be raised over the infra-red searchlight to prevent crushing it. Hatches above the troop compartment cannot open while the gun is aimed backwards. The turret is challenging to aim because of its low profile. On the left side of the hull, the driver is seated in the front. When his hatch is closed, he has three periscope vision blocks to help him see. When swimming, the driver's centre vision block can be swapped for an extended periscope or active night-vision binocular equipment for use in low light. The BMP was the first Soviet armoured vehicle to employ a basic yoke steering system. With an effective range of around 400 metres (440 yards), the commander's station, which is situated behind the driver's station, is equipped with a detachable infrared searchlight.

Iraq had BMP-1s left over from the previous conflicts and used them against Coalition forces in the initial fighting in 2003. These had been purchased from the Soviet Union in 1980 in standard green, after which they would be transformed into the Iraqi Army's desert paint scheme. (DPL)

A shot of the BMP-1's two separate rear doors leading to compartments for personnel. The team commander sat towards the front and could speak to the driver; his positioned steering yoke can be seen in the left side of the rear passenger area. (US DOD)

BMP-2 Armoured Personnel Carrier

Developed in the 1980s, the BMP-2 offered increased mobility and firepower to the military forces using it. It was highly popular with Iraqi troops, particularly as the crew benefited from an air-conditioning system – when it worked. The vehicle carries three crew members, including a driver, gunner, and commander as well as seven soldiers in the troop compartment. One of the key differences on the new

Right: A damaged and abandoned Iraqi BMP-2K armoured command vehicle sits along a roadside in northern Iraq during the 2003 invasion. (US DoD)

Below: Iraq went on to purchase the BMP-2 from the Soviet Union. These were regarded as very reliable when first delivered, but a lack of maintenance and spare parts caused serviceability issues. (Ministry of Defence of the Russian Federation)

BMP-2 was that the turret was larger than that of the BMP-1; this allowed for both the commander and gunner to be positioned inside, whereas only the gunner could be carried in the BMP-1 turret. A 30mm AG-17 automatic grenade launcher (AGL), a 7.62mm PKT coaxial machine gun and a 30mm 2A42 automatic gun are also mounted on the turret. The 7.62mm gun can deliver 2,000 rounds per minute, the 30mm cannon can fire 500 rounds per minute and the AGL, with a maximum range of 1,730 metres, can launch 250 grenades per minute. According to the designers, the KBP Instrument Design Bureau, the vehicle can withstand high-explosive anti-tank (HEAT) missiles and 12.7mm B-3232 rounds, thanks to the appliqué armour protecting its hull. At 6.73 metres long, 3.15 metres wide and 2.45 metres high, it maintains the dimensions of the original BMP-1 design parameters.

BMP – Armoured Infantry Fighting Vehicle Specifications		
Model	BMP1 IFV (Soviet Union)	BMP 2 IFV (Soviet Union)
Manufacturer	Kurganmashzavod (Soviet Union) and ZTS Detva Designed by Pavel Isakov Design Bureau	Kurganmashzavod, Ordnance Factory Medak
Country	Iraq	Iraq
Year	1966–present	1979–present
Engine	UTD-20 V6 diesel engine 300hp (224kW) at 2,600rpm	diesel UTD-20/3 300hp
Fuel	Diesel	Diesel
Protection	Welded rolled steel (33mm)	33mm
Top Speed	40mph (64km/h)	40mph (64km/h)
Range	350 miles (563km)	370 miles (595km)
Crew Capacity	3 plus 8 passengers	3 plus 7 passengers
Length	22.1ft (6.735m)	
Width	9.8ft (2.94m)	10.4ft (3.15m)
Height	6.9ft (2.068m)	8ft (2.45m)
Armament	One 73mm 2A28 smoothbore gun or 9M14 Malyutka ATGM	30mm 2A42 autocannon, 9M113 Konkurs ATGM
Weight	14.5 tons (13.2 tonnes)	15.7 tons (14.3 tonnes)
Service Branch	Iraqi Army	

MT-LB Personnel Carrier

The Soviet-produced multi-purpose, fully amphibious, tracked armoured fighting vehicle known as the MT-LB has been in service since the 1970s and made up a significant proportion of Iraq's armoured capability in 2003.

This armoured amphibious platform can carry 11 fully equipped soldiers. The driver and the commander/gunner, who make up the crew, are seated in a compartment at the front of the car, with the engine situated behind them. Up to 4,400lb of equipment can be carried in the back. In addition, the MT-LB can tow up to 6,500kg (14,300lb) of cargo. The vehicle is entirely amphibious, moving through the water on its tracks. The vehicle has a maximum thickness of 14mm (0.55in) of steel armour for the turret front and 3–10mm (0.12–0.39in) against small-arms fire and shell splinters.

Jubilant Iraqi soldiers riding on an MT-LB on a highway in Iraq in February 2003. The MT-LB was a basic personnel carrier designed and built by the Soviets and sold to Iraq. (DPL)

The remains of an MT-LB multi-purpose armoured vehicle with a ZU-23-2 anti-aircraft gun fitted. This example was hit by Coalition armour near Baghdad. (DPL)

Its primary armament options include a potent 12.7mm NSV/Kord heavy machine gun or a 30mm AGS17D/AGS-30 automatic grenade launcher, or alternatively a 30mm 2A42/2A72 autocannon. Supporting its offensive capabilities, the secondary armament comprises a PKT with a generous ammunition capacity of 2,500 rounds.

Propelled by a YaMZ 238 V-8 diesel engine, producing 240hp at 2,100rpm, or an SW 680 I6 diesel engine with the same power output at 2,200rpm (especially employed in Poland), the MT-LB exhibits a power-to-weight ratio of 20hp/tonne. It utilises a torsion-bar suspension system. Demonstrating an operational range of 500km (310mi) on roads, it can achieve a maximum speed of 61km/h (38mph) on such surfaces. Off-road capabilities are slightly reduced to 30km/h (19mph), while in aquatic environments, it can navigate at a speed of 5–6km/h (3.7mph). Iraq upgraded its MT-LBs to become self-propelled anti-aircraft guns (SPAAGs) by equipping the rear of the vehicle with a ZU-23-2 23x152mm twin anti-aircraft gun. The gun's wheels were taken off during this process; thus, it is difficult to detach and use it separately. This particular adaptation came in at least two variations: one had the ZU-23-2 positioned in an open-topped turret, while the other had it mounted on a platform that extended past the MT-LB's hull and had a roof over the gun operators. The second version's roof would obstruct the gun's sights at a high altitude, suggesting it was designed primarily for use as fire support.

MT-LB Armoured Personnel Carrier Specification	
Model	MT-LB
Manufacturer	Kharkov Tractor Centre, USSR
Country	Iraq
Year	1950s–80s
Engine	YaMZ 238 V8 Diesel
Diesel	Diesel
Protection	Welded rolled steel (14mm)
Top Speed	38mph (61km/h)
Range	310 miles (498km)
Crew Capacity	2 plus 11 passengers
Length	21.2ft (6.45m)
Width	9.5ft (2.86m)
Height	6.1ft (1.86m)
Armament	12.7mm NSV/KOrd heavy machine gun
Weight	13.1 tons (11.9 tonnes)
Service Branch	Iraqi Army

Panhard AML-60/90

The Panhard AML reconnaissance armoured car was a French vehicle that Iraq had procured long before the invasion of Kuwait. Designed on a 4×4 chassis and weighing approximately five tons, the Panhard was deployed in small numbers in 2003, with many being held back in Baghdad.

Introduced in 1959, AMLs have been used globally, with variants remaining in production for more than 50 years. Its forerunner, the AML, was once considered among the most heavily armed scout vehicles, featuring a 90mm rifled cannon, a 60mm breech loading mortar and dual 7.5mm machine guns in the turret. Capable of engaging targets at 1,500 metres, it was a formidable force against second-line and older main battle tanks. The AML features coil spring suspension and drum brakes, but lacks hydraulic assist for brakes and steering. Steering, controlled only by the front wheels, requires considerable effort while in motion and remains effectively locked when stationary. Rear-wheel drive is transmitted directly to epicyclic hub reduction gears. The vehicle employs a centrifugal clutch with electromagnetic control, eliminating the need for a pedal. The gearbox consists of separate high and low-range gearboxes, each serving specific terrain conditions.

The Panhard AML reconnaissance armoured car was a French vehicle that Iraq had procured long before the invasion of Kuwait. Designed on a 4×4 chassis and weighing approximately five tons, the Panhard was deployed in small numbers in 2003, with many being held back in Baghdad. (US DoD)

Left: The Panhard's rear enclosed powerplant was shielded from attack. Its hull was assembled from 13 welded pieces, but its turret basket was cramped, with limited space due to the gun breech. The turret has a two-man crew, with the commander on the left and gunner on the right. (US DoD)

Below: In 2003, Iraq deployed the EE-9 Cascavel. The vehicle is designed to deflect and counter missile and rocket attacks with a steep frontal glacis that slopes upwards and back towards the horizontal hull roof. The Brazilian-produced vehicle was retained in service after the 2003 war. (George Williams)

The AML's powerplant, a 1.99L four-cylinder engine inspired by the Panhard EBR, is air-cooled but somewhat underpowered for its weight. The hull is assembled from 13 welded pieces, featuring a cramped turret basket and limited space due to the gun breech. The turret has a two-man crew, with the commander on the left and gunner on the right. Access doors on either side, sand channels for obstacles and nitrogen inner tubes with run-flat capability enhance the vehicle's functionality.

In Iraqi service during the 1991 Gulf War, AML platoons, primarily comprising AML-90s, were attached at the brigade or battalion level for reconnaissance. An Iraqi armoured reconnaissance platoon could consist of up to 8 AMLs. The AML-90s were valued for their armament size and range, while AML-60s had secondary roles. In the Battle of Khafji in 1991, AML-90s were deployed ineffectively against US Marine Corps and Saudi National Guard troops. Iraqi crews often failed to utilise the vehicles' mobility and were engaged from static positions, resulting in significant losses. The US estimated that Iraq operated 300 AMLs in 1990, losing about half of them during the Gulf War. Despite some remaining in service in 2003, Iraqi AMLs faced challenges due to erratic maintenance and a lack of spare parts. They clashed with American tanks during the invasion of Iraq in 2003 in the vicinity of Nasiriyah.

Panhard AML – Armoured Reconnaissance Vehicle (Amphibious Capable) Specifications		
Model	Panhard AML 60	Panhard AML 90
Manufacturer	Panhard (France)	
Country	Iraq	
Year	1960–87	
Engine	Panhard 1.99L Model 4 HD flat 4-cylinder air-cooled petrol	Panhard gasoline, 1.99L Model 4 HD 4-cylinder water-cooled 90hp
Fuel	Petrol	
Protection	Welded rolled steel (33mm)	18mm steel
Top Speed	62mph (99km/h)	65mph (104km/h)
Range	370 miles	250 miles
Crew Capacity	3 (commander, driver, gunner)	
Length	12.4ft (3.79m)	16.7ft (5.11m)
Width	6.46ft (1.97m)	6.46ft (1.97m)
Height	6.79ft (2.07m)	6.79ft (2.07m)
Armament	60mm HB 60 breech-loaded mortar	90mm D921 F1 rifled cannon
Secondary Armament	20mm M621 autocannon	7.5mm AAT-52 or 7.62mm MAG machine gun
Weight	5.2 tons (4.8 tonnes)	6 tons (5.5 tonnes)
Service Branch	Iraqi Army	

EE-9 Cascavel

The EE-9 Cascavel is a Brazilian-made armoured vehicle. The driver's seat is positioned at the front left, a central turret (manual in Mk II, electrically powered in later variants), and the motor and transmission at the rear. The Cascavel Mk III is armed with an Engesa EC-90 90mm gun capable of firing various shells, accompanied by a coaxial 7.62mm machine gun. The EC-90 has limited elevation and depression angles, lacks stabilisation, and has a basic optical fire-control system, upgraded with a laser rangefinder in Brazilian service. Late-production Cascavels include run-flat tyres and a unique central tyre-pressure regulator accessible from the driver's compartment.

The EE-9 Cascavel is characterised by its boxy, boat-shaped design. The vehicle features a steep frontal glacis that slopes upwards and back towards the horizontal hull roof, incorporating recesses for headlamps and a thick glacis plate over the driver's seat. The nearly vertical hull sides are also sloped inwards towards the roof. At the forward section of the hull, there is a low, well-rounded turret housing a long, tapered gun barrel and a triple-baffle muzzle brake. Many armies have taken a liking to the EE-9 Cascavel because of its straightforward design and use of parts that are already widely used in civilian industry. [35] Developing countries find it particularly appealing to acquire due to its affordable price when compared to similar Western armoured automobiles. The purely commercial nature of Engesa sales, free from any political supplier constraints, was also seen as a respectable substitute for weapons from the Warsaw Pact and NATO during the height of the Cold War.

Early in the 1980s, the Cascavel III was developed to give the Cascavel a more potent punch than the low-pressure 90mm (3.54in) DEFA rifle. Engesa created a new turret for the Cascavel III. This turret had a Belgian Cockerill 90mm (3.54in) gun mounted on the same turret ring. A coaxial 7.62mm (0.3in) Browning LMG or Belgian FN Mag was used as the secondary armament. On the exterior pintle mount, the Browning M2HB 12.7mm (0.5in) heavy machine gun could be employed. The engine was changed to a 6-cylinder, water-cooled Detroit Diesel 6V-53N with 212 horsepower. It enabled the car to reach a maximum speed of 60mph (100km/h) and a total range of 500 miles (800km). Inside the front armour were fitted headlights and road lights. In terms of exports, this model was most likely the most successful of all the Cascavel versions. Specifically, the gun improvement provided much better anti-tank performance. The muzzle velocity speed of the MECAR Kenerga 90/46 and M603 rounds was 1,430m/s.

EE-9 Cascavel – Armoured Reconnaissance Vehicle Specification	
Model	EE-9 Cascavel
Manufacturer	Engesa | Brazil
Country	Iraq
Year	1974–93
Engine	Detroit Diesel 6-cylinder water-cooled. Automatic transmission
Fuel	Petrol
Protection	Welded rolled steel (33mm)
Top Speed	62mph (99km/h)
Range	470 miles (756km)
Crew Capacity	3
Length	20.8ft (6.29m)
Width	8.6ft (2.59m)
Height	8.6ft (2.59m)
Armament	90mm Engesa EC-90
Weight	13.2 tons (12 tonnes)
Service Branch	Iraqi Army

WZT-2 Armoured Recovery Vehicle

Based on the Soviet T-55 hull, the 34-tonne WZT-2 was a Polish-built Armoured Recovery Vehicle (ARV) designed to recover the T-55. It was equipped with a crane to carry out engine changes and a winch to haul stranded tanks. Fitted with one weapon, a 12.7mm DShK heavy machine gun, the WZT-2 was able to carry out repairs on the T-54/55 and T-72 and has the ability to tow vehicles up to 40 tonnes. It has the ability to cross shallow rivers and can transport a small number of wounded soldiers in emergency situations. The crew of three were able to use a dozer blade fitted to the WZT-2.

The platform supported Iraq's sizeable fleet of battle tanks during both major wars. During the Gulf War of 1991, many of the WZT-2 recovery vehicles were destroyed, but a unit of 15 remained in service in 2003. After Operation *Iraqi Freedom*, a handful remained in service, with most of those seen in Baghdad in need of repair themselves.

WZT-2 Armoured Recovery Vehicle Specification	
Model	WZT
Manufacturer	Bumar-Łabędy (Poland)
Country	Iraq
Year	1973–present
Engine	V-55A 12-cyl. 38.88L water-cooled 591hp diesel
Fuel	Diesel
Protection	Hull 20–79mm steel armour
Top Speed	31mph (50km/h) road, 16mph (27km/h) off road, 9mph (15km/h) towing
Range	288 miles (465km)
Crew Capacity	4 crew plus 3 wounded
Length	21.16ft (6.45m)
Width	10.72ft (3.27m)
Height	6.88ft (2.1m)
Main Armament	12.7mm DShK
Weight	37.4 tons (34 tonnes)
Service Branch	Iraqi Army

The WZT-2 was Iraq's armoured repair and recovery vehicle, deployed to support T-72s, T-62s and other tanks. (US DoD)

Chapter 4
The Multi-National Force (MNF)

Coalition Joint Task Force 7 – Multi-National Force
International concern about the invasion of Iraq had made many governments cautious; Germany and France, among others, had objected to the invasion and declined to send military forces. Once the initial invasion phase was completed and US and UK troops secured Baghdad and Basra, many countries offered to send troops to support a peacekeeping operation across the country in a US-led Multi-National Force, co-ordinated from Baghdad. More than 40 nations stepped forward, among them Australia, Denmark, Italy, Romania, Poland and Ukraine, deploying armoured units to support what became an enduring campaign.

The Multi-National Force established itself in the Green Zone, a fortified section of Baghdad city centre previously administered by the Iraqi government. The command group was codenamed Coalition Joint Task Force 7 (CJTF7).

Many of the troops involved in the invasion remained in theatre for this new role, constituting the armoured spearhead of the USA and UK, while Australia and Poland had provided ground troops for specialist tasks in southern and western Iraq. In the north, the Kurdish Peshmerga supported the Coalition and operated in support of US forces. In this way, the administration of Iraq was split into areas supervised by a particular armoured contingent, with Polish forces in the west, Italians in the central region and the British in the south, while the US oversaw operations in Baghdad, the north and east. Commanded by the CJTF7 headquarters in Baghdad, these regions were listed as Multi-National Division West (MNDW), Multi-National Division North (MNDN), Multi-National Division South (MNDS) and Multi-National Division South East (MNDSE). In the wake of the invasion and with major combat operations ending, the US formed an interim administration called the Coalition Provisional Authority (CPA), which governed the country in the short term.

The jubilation among the many Iraqi Shias at being liberated from the tyranny of Saddam was evident everywhere, but it quickly faded as the head of the CPA, Paul Bremer, ordered the disbanding of the 300,000-strong Iraqi Army and its replacement by a new force, called the New Iraqi Army, to be raised and trained by the Coalition. With Iraq's power and water supplies damaged in air strikes and still not repaired, food was in short supply and the act of making thousands of soldiers redundant left families with no income. These former soldiers now joined the growing insurgency and directed their anger at the United States. Within months, attacks on Coalition forces soared and the security situation deteriorated. The vast ammunition dumps left unguarded across the country were ransacked and used to make Improvised Explosive Devices (IEDs). Hundreds of Coalition soldiers were injured in roadside attacks and by the end of 2007, senior commanders estimated that 63 per cent of Coalition deaths were due to IEDs. In a move to offer some protection, the United States provided armoured Humvees to many countries that had no protected vehicles, while others shipped in their own armour. Washington and London simultaneously ordered new bespoke protected vehicles for their troops.

Almost all the contributing forces, from Azerbaijan to Hungary, South Korea, Estonia, the Netherlands, Australia, Slovakia, Georgia, El Salvador, Denmark, Spain, Ukraine, Bulgaria, Italy and Poland lost soldiers, with the United States and Britain recording the largest number of casualties. Though more armour appeared on the streets to protect soldiers, the policymakers failed to understand the prevailing culture and would achieve comparatively little in the struggle to train and develop a New Iraqi Army.

Australia

In 2003 Australian premier John Howard and his Cabinet made the decision to commit Australian troops to the US-led military intervention in Iraq. It initially sent a ground force, operating in the western desert, and later provided a substantial contingent under the codename Operation *Falconer*, in which it deployed air, naval and land assets. However, the Australian government refused US and UN requests made later in 2003 and into 2004 to deploy additional forces. In 2004, Canberra argued that as well as its forces on the ground, the Australian military had trained and prepared a Fijian force for deployment to Iraq and had therefore more than met its obligation to contribute to security arrangements. Then, in February 2005, the administration announced that the Australian Army would deploy a battlegroup to Al Muthanna Province in south-west Iraq, to provide security for Japanese engineers deployed to the area as well as to help train Iraqi personnel of the New Iraqi Army. The resulting Al Muthanna Task Group (AMTG), which was approximately 500 strong, was equipped with armoured vehicles including the Australian Light Armoured Vehicle (ASLAV) and the armoured Bushmaster. Following the withdrawal of the Japanese force and the transition of Al Muthanna to Iraqi control, the Australian battlegroup relocated to Tallil Air Base in neighbouring Dhi Qar province in July 2006.

The name AMTG was subsequently abandoned in favour of the title Overwatch Battle Group (West), reflecting the unit's new role. Al Muthana and Dhi Qar are the westernmost of Iraq's four southern provinces and OBG(W) became the prime coalition intervention force in the western sector of the British-controlled Multi-National Division South East (MNDSE) area of operations. Using ASLAVs and Bushmasters, this unit was able to cover vast areas of desert. By late 2006, overall personnel numbers committed to the operation had risen to 1,400. Then, after the Labour Party achieved a landslide victory in the Australian election of 2007, the new Prime Minister, Kevin Rudd, announced the withdrawal of Australian combat forces from Iraq, which began on 1 June 2008. The OBG(W) and Australian Army Training Team formally ceased combat operations on 2 June, having helped train 33,000 Iraqi soldiers.

ASLAV (Australian Light Armoured Vehicle)

The ASLAV served as Australia's primary reconnaissance and armoured patrol vehicle in Iraq, offering mobility and protection for reconnaissance tasks. As part of its commitment to the Iraq War, the Australian Army deployed numerous units to serve on Overwatch Battle Group OBG(W) rotations in the Dhi Qar Governorate, in addition to the Security Detachment (SECDET) task in Baghdad. In 2004, an ASLAV from the Security Detachment engaged and destroyed an insurgent mortar baseplate, which was launching projectiles into the Green Zone in Baghdad. This marked the first Australian engagement of the ASLAV's 25mm main gun against hostile troops. The ASLA can be fitted with enhanced armour and can counter small arms and rocket attacks.

In upgrades for Iraq, the vehicle was equipped with run-flat tyres and air conditioning, with wide wheels and tyres fitted to cope with the desert terrain. Because of its dependability, cheap maintenance costs and capacity to move swiftly across great distances, the ASLAV was constantly deployed on operations. Just some of its advanced equipment are an integrated laser range finder, an improved drivetrain, better thermal optics and an electric turret. Multi-system surveillance suites and remote weapon stations are also installed in some vehicles. Inside is a suite of countermeasures, including against electronic and improvised explosive

Before its deployment to Iraq, the Australian military spent several months training in Darwin. Once in Iraq, the ASLAVs received armour and electronic warfare enhancements. (Australian DoD)

The Australian Security Detachment (SECDET) outside its embassy in the Green Zone. The team used ASLAVs and Bushmasters in the role of escorting and protecting diplomatic staff and visitors to Baghdad. (Australian DoD)

devices, as well as a fire suppression system. The use of non-permanent mission role installation kits and extra armour, has further expanded the ASLAV's adaptability. Other variants of the ASLAV include an ambulance, a fitter and recovery platform, a command vehicle, armoured personnel transport, surveillance and reconnaissance. The vehicle can carry nine crew with full equipment. The cost of preparing ASLAV crews for combat and training them is greatly decreased by the use of simulators.

The Australian Army plans to replace the ASLAV with the eight-wheeled Boxer combat reconnaissance vehicle, which is expected to enter service in 2026.

ASLAV (Australian Light Armoured Vehicle) Specification	
Model	ASLAV
Manufacturer	General Motors Diesel & General Dynamics Land Systems Australia
Country	Australia
Year	1995–present
Engine	Detroit Diesel 6V-53T 205kW (275hp)
Fuel	Diesel
Protection	Steel armour
Top Speed	62mph (100km/h)
Range	410 miles (660km)
Crew Capacity	3 plus 6 passengers
Length	21.4ft (6.53m)
Width	8.6ft (2.62m)
Height	8.8ft (2.69m)
Main Armament	M242 25mm Bushmaster chain gun
Secondary Armament	2 × 7.62mm MAG58 machine gun
Weight	14.5 tons (13.2 tonnes)
Service Branch	Australian Army

The ASLAV's high wheelbase and angled chassis aided in protection from roadside blasts. In this image, Bushmaster armoured vehicles can be seen in the background. (Australian DoD)

In Iraq, the highly versatile ASLAV, which also has an amphibious capability, was used with the driver's cupola closed to avoid any sniper or grenade attacks. (Australian DoD)

After the increase in insurgent attacks on Coalition forces, Australian commanders made a decision to send the Bushmaster vehicle to Iraq. It carried ten soldiers and quickly earned a reputation as almost 'bulletproof' due to its armoured protection. (Australian DoD)

Bushmaster Protected Mobility Vehicle

The increase in insurgent attacks on Coalition forces in 2003 prompted a decision by Australian commanders to send additional protected vehicles to Iraq. The Bushmaster Protected Mobility Vehicle (PMV-M), carrying ten soldiers, was deployed and was an immediate success due to its special design, by which its armoured V-shaped hull shields its occupants from landmines and other explosives. Blasts are deflected upward and away from the vehicle by the hull's sloped sides and similarly shaped windows. The vehicle's single-piece, welded shell is intended to ward off all small-arms fire. To shield soldiers from potential fires, the fuel and hydraulic tanks of the Bushmaster are situated outside the crew's compartment. In order to prevent the car from becoming stranded, there is also a shielded emergency petrol tank. As well as in Iraq, the Bushmaster has continued to be used by Australian forces as well as by the Dutch; in a number of incidents vehicles have been blown up but the soldiers remained unharmed.

Bushmaster Protected Mobility Vehicle (Bushmaster PMV) Specification	
Model	Bushmaster
Manufacturer	Thales Australia
Country	Australia
Year	1997–present
Engine	Caterpillar 3126E 7.2L 6-cylinder diesel, turbocharged 224kW (300hp)
Fuel	Diesel
Protection	High Hardness Steel
Top Speed	62mph (100km/h)
Range	497 miles (800km)
Crew Capacity	10 (1 driver plus 9 passengers)
Length	23.5ft (7.18m)
Width	8.12ft (2.48m)
Height	8.69ft (2.65m)
Main Armament	Remote weapon station up to 12.7mm HMG or 40mm AGL, or manned open turret up to 12.7mm HMG or 40mm AGL
Secondary Armament	Manned swing mounts up to 7.62mm (one front and two rear)
Weight	16.5 tons (15 tonnes)
Service Branch	Australian Army

Poland

On 17 March 2003, Polish President Aleksander Kwaśniewski announced that his country would send 600 troops to the Persian Gulf to take part in the invasion of Iraq; this number was subsequently increased to 2,000. Polish commandos took part in security operations to secure Iraqi oil platforms, after fears that Saddam's forces would set them alight, as they had done in 1991. On land, the commandos used the Humvee as their main protected mobility, which was upgraded with US enhanced armour packages as the IED threat increased. Shortly after the Polish contingent's arrival in Iraq, a political row erupted between Poland and France after Polish forces discovered French-made Roland short-range surface-to-air missiles, which it was claimed had been manufactured in 2003 and sent to Iraq. France pointed out that the latest Roland missiles had been manufactured in the early 1990s and thus the manufacturing date on the missile was an error.

Armoured Humvee

Poland deployed the Humvee to Iraq in 2003, having purchased the armoured platform from the United States. Loudspeakers were mounted on the vehicles to deliver messages to the local community using Iraqi interpreters.

The Humvee underwent major modifications in late 2003 as the role of American forces in Iraq changed from fighting the Iraqi Army to suppressing an insurgency of thousands of fighters across the country. On the main road between the Green Zone, an area in central Baghdad where the US headquarters was based, and the airport, youngsters were paid $10 by insurgents to drop grenades on to Humvees. Most bounced off and exploded in mid-air or hit innocent civilians in a car behind, but occasionally they injured or killed a soldier. Elsewhere, insurgents would pose as market traders and pack their trolley with fruit hiding an IED, which they would then abandon as a military patrol came close. To counter these attacks, armoured kits were manufactured and military scientists worked on further enhancements. While these kits are much more effective against all types of attacks, they weighed anything from 1,500–2,200lbs (680–1,000kg) and risked compromising the vehicles' capability. Most up-armoured Humvees held up well against lateral attacks when the blast was distributed in all different directions but offered little protection from a mine blast below the truck, such as from buried IEDs, landmines and explosive formed penetrators.

The majority of Humvees used by Polish forces had undergone the major armoured upgrades, being equipped with the Armour Survivability Kit (ASK) as well as the FRAG 5 and FRAG 6 upgrades to the Humvee M1151 variant. The ASK kit was first employed in October 2003, adding about 1,000lb (450kg)

Polish soldiers prepare to embark on a patrol in their upgraded armoured Humvees. The extra thickness of the armoured doors can clearly be seen. 2003 (Polish MoD)

to the weight of the vehicle. A lighter kit added to subsequent vehicles had a penalty of just 750lb (340kg). From January 2005, the US Marines developed their own improvements, called the Marine Armour Kit (MAK), which offered more protection but also increased weight. The FRAG 5, for example, offered blast protection but was still inadequate to stop Explosive Formed Penetration (EFP) attacks. The FRAG 6 kit was designed to do just that, but its increased protection added more than 1,000lb (450kg) to the vehicle and increased its width by 2ft (61cm). The doors also now required a mechanical assist device to open and close due to the weight. Mechanical upgrades were also incorporated, the final stages of these making the Humvee highly capable against IED and rocket-propelled grenade attacks.

Humvee – High Mobility Multipurpose Wheeled Vehicle (HMMWV) UAH Specification	
Model	M1114 UAH
Manufacturer	AM General
Country	United States, purchased by Poland
Year	1984–present
Engine	6.2L V8 diesel or 5.7L gasoline or 6.5L V8 turbodiesel and non-turbo diesel: 190hp
Fuel	Diesel
Protection	Welded aluminium, composite and steel armour protection (varies on armour package)
Top Speed	55mph at max gross weight. Over 70mph top speed
Range	300 miles (480km)
Crew Capacity	4; varies on configuration
Length	15ft (3.3m)
Width	7.1ft (2.16m)
Height	6ft (1.37m)
Main Armament	Multiple configuration. Typically, M2 HMG, M134 Minigun or Mk19 Grenade Launcher
Weight	6.6 tons (6 tonnes)
Service Branch	Polish Army

A Polish Humvee operating with the Civil and Military Cooperation (CIMIC) team near Al Diwaniyah, east of Najaf, during Operation *Iraqi Freedom*. (US DoD)

Heavily armoured Polish Humvees. The enhanced gun position included armoured glass to allow soldiers to pinpoint the enemy from a protected position. (US DoD)

Polish Humvees with enhanced armour at a forward operating base in Iraq. All the vehicles bear the Polish flag. (US DoD)

Denmark

In March 2003, the Danish Parliament authorised the country's support of the US invasion of Iraq and agreed to the active military participation of Danish forces. The government in Copenhagen approved the deployment of a submarine and a corvette off southern Iraq, and a military medical unit as well as armoured units to support the land operation. As such, Denmark became one of a small handful of nations, and the only Nordic state, to participate militarily in the initial invasion of Iraq. After the invasion phase, Denmark agreed to broaden its support for the occupation and nation-building mission and deployed troops. These were attached to the British in Multi-National Division South East (MNDSE), which included Basra, Maysan, Dhi Qar and Al Muthanna provinces. Dancon, the Danish force, arrived in Kuwait in the first week of June 2003, and reached its area of operation in Iraq on 6 June. After an initial set-up phase, the unit commenced operations on 12 June, replacing the British Chemical Warfare Regiment at Al-Qurna. The formation was initially 380 personnel strong, with 42 more added in July and August. By February 2005, the contingent consisted of approximately 545 soldiers. Danish troops were rotated every six months, and each contingent had a slightly different composition of units. By March 2007, its size had fallen to about 460. There was also a 53-man Lithuanian unit, designated Litcon (Lithuanian Contingent), which had been attached to Dancon since June 2003. It was withdrawn along with the Danes by August 2007.

MOWAG Piranha

The Mowag Piranha series of wheeled armoured vehicles was designed by Mowag, a Swiss company later acquired by General Dynamics. Denmark procured more than 100 Piranhas in the armoured personnel carrier role as well as anti-tank, command post, ambulance and communications platform variants, in support of Tactical Air Control Parties' (TACP) co-ordination of Close Air Support (CAS) from fighter aircraft.

In March 2003, the Danish Parliament authorised the country's support of the US invasion of Iraq and agreed to the active military participation of Danish forces. The eight-wheeled variant of the MOWAG Piranha was deployed to Iraq, fitted with enhanced armour to protect from rocket-propelled grenades and roadside bombs. (Danish MoD)

Since its introduction in 1972, this incredibly adaptable modular vehicle has been manufactured in four variants, which are; 4x4, 6x6, 8x8, and even 10x10. The MOWAG Piranha III series' hull is made entirely of welded, High-Hardness Steel armour, shielding its occupants from shell splinters and small-arms fire. Based on a revolutionary lightweight hull design with improved variable ballistic protection and larger payload, the third-generation Piranha family boasts superior performance. In 2003, the Danes deployed the eight-wheeled variant of the MOWAG Piranha to Iraq, fitting it with enhanced armour for protection from rocket-propelled grenades and roadside bombs.

Mowag Piranha Armoured Personnel Carrier Specification	
Model	Piranha III
Manufacturer	MOWAG
Country	Switzerland
Year	1972–present
Engine	Diesel 202kW 275hp engine
Fuel	Diesel
Protection	Quick-mount selection, including defence against RPG, IED, NBC
Top Speed	62mph (100km/h)
Range	485 miles (780km)
Crew Capacity	3 (driver, commander, gunner) plus 5 passengers
Length	20.6–24.5ft (6.25–7.45m)
Width	8.2–8.9ft (2.5–2.66m)
Height	5.11–6.6ft (1.8–1.98m)
Main Armament	1 × 12.7mm MG turret, or MOWAG apex mount, grenade launcher, or TOW anti-tank missile
Weight	10.2 tons (9.3 tonnes)
Service Branch	Danish Army

The Piranha's forward hatch had several settings, allowing it to be driven slightly open or fully closed. Denmark procured more than 100 Piranhas in various variants in support of Tactical Air Control Parties (TACP) to co-ordinate Close Air Support (CAS) from fighter aircraft. (Danish MoD)

Italy

In April 2003, the Italian government approved the mobilisation of forces to support operations in Iraq, with a planned deployment in July 2003. Under the codename Operation *Ancient Babylon*, the Italians were based at Nasiriyah in the south of the country and assisted by Portuguese and Romanian troops.

In November 2003, Italy lost 36 soldiers in a bomb attack that impacted upon public support for the country's role in the war. Then, between April and August 2004 in Nasiriyah, Italian troops came under fierce attack from the Mahdi Army in a campaign known as the Battle of the Bridges. Casualties were high and in the ongoing firefights, the Italians' eight-wheeled Centauro howitzer played a major role in defeating the enemy.

On the night of 6 April, about 500 Italian soldiers faced a force in excess of 1,000 militiamen at the objective close to all three bridges, but as a result of women and children gathering among the militiamen on the third bridge, the Italians did not take any action to cross it. In a second incident subsequently called the Porta Pia operation, various companies from the Italian force were engaged, including the 11th Bersaglieri regiment, a company from the San Marco battalion, a heavy armoured squadron and the Centauro del Savoia cavalry, as well as the GIS carabinieri and paratroopers of the Tuscania regiment. During the battle, the Italian military was targeted with more than 400 anti-tank rockets. The Italians fired five or six Milan missiles to neutralise four enemy positions, but a deluge of fire stopped their advance. The eight-wheeled Italian Centauro was at the forefront of their offensive and fought a hard battle. As a consequence of their experience, the Italians decided they needed heavy armour. In 2006, the Italians deployed a small number of Ariete tanks, which were not used in anger.

Ariete MBT

In 2006, Italy deployed a small number of Ariete Main Battle Tanks to provide surveillance, enhanced firepower and deliver a psychological message of power and intent. The Arietes, however, were never deployed in action but were maintained as a mobile reserve.

The Ariete is a third-generation main battle tank developed by Consorzio Iveco Oto Melara (CIO), a consortium formed between Iveco (which produced the chassis and engine) and OTO Melara (which supplied the turret and fire control system). The vehicle carries the latest optical and digital imaging and fire-control systems, enabling it to fight by day and night and to fire on the move. Deliveries were first planned for 1993, but didn't take place until 1994 due to delays. The Ariete's main armament is a 120mm smoothbore gun, similar to the Rheinmetall L/44. It fires most NATO-standard rounds of this calibre and the crew carries 42 rounds, 15 of which are stored vertically on the left side of the main gun breech. The other 27 are stowed in a hull rack to the left of the driver's station. The gun barrel has a thermal

The Italian Ariete Main Battle Tank, which was sent to Iraq during 2006 to provide enhanced firepower to the Italian force. It delivered a psychological message of power and intent. The Ariete was never deployed in action, however, and was maintained in a mobile reserve. (Italian MoD)

insulating sleeve and a fume extractor, and is fully stabilised in both azimuth and elevation by an electro-hydraulic drive system. In addition to having four forward and two reverse speeds, the fully automatic gearbox system includes a hydraulic retarder and steering system.

The tank enjoys more mobility in all operating circumstances thanks to the expanded track system and enhanced heavy-duty final drives. Four return rollers on each side and seven twin rubber-lined road wheels make up the running gear. Each suspension arm has a hydraulic bumper and a torsion bar as part of the suspension system. In order to meet the enhanced performance, the brake system has also been modified. In addition to a 127mm machine gun for air and local defence, a 7.62mm machine gun, coaxially connected to the main armament weapon of the same calibre, is installed on the turret roof.

Ariete Main Battle Tank Specification	
Model	Ariete C1
Manufacturer	Consorzio Iveco OTO Melara (CIO)
Country	Italy
Year	1994–present
Engine	Fiat MTCA 12V diesel 25.8L 950kW (1,270hp)
Fuel	Diesel
Protection	Steel and composite
Top Speed	40mph (65km/h)
Range	370 miles (600km)
Crew Capacity	4 (commander, driver, gunner, loader)
Length	31.2ft (9.52m)
Width	11.8ft (3.61m)
Height	8ft (2.45m)
Main Armament	120mm OTO Breda L/44 smoothbore gun. 42 rounds
Weight	59.5 tons (54 tonnes)
Service Branch	Italian Army

The Ariete's main armament is a 120mm smoothbore gun, similar to the Rheinmetall L/44. The gun fires most NATO-standard rounds of the same calibre. The crew carries 42 rounds, with 15 ready rounds stored vertically on the left side of the main gun breech. The other 27 are stowed in a hull rack to the left of the driver's station. (Italian MoD)

Dardo Infantry Fighting Vehicle

Six Dardo IFVs were deployed to Iraq in 2003 with the Italian forces, providing armoured transport and fire support for infantry units.

Produced in Italy, the Dardo was a mechanised infantry fighting vehicle and was introduced to support the M113, which was in Italian service in the beginning of the 1980s. The engine and driver are located in the front of the Dardo's hull, following a traditional layout by which the crew compartment is located at the back and the two-man turret is positioned in the centre. The ramp with the rear door opens for the seven soldiers to enter the vehicle. The infantry personnel variant was mass-produced out of the many variations that were suggested.

The hull of the Dardo is made of aluminium and has extra steel armoured plates on top. This offers complete protection across the frontal arc against heavy calibre 14.5mm bullets, small weapons fire and shell splinters. Two banks of four smoke grenade launchers are installed, along with a protective nuclear, biological and chemical (NBC) warfare system. In addition, the Dardo was fitted in Iraq with an appliqué armour package that added three tonnes to the weight of the vehicle. Over the frontal arc, the appliqué armour provides increased protection against 25mm APDS bullets.

The Dardo infantry fighting vehicle was deployed to Iraq in 2003 with the Italian force. A total of six Dardo vehicles served in Iraq, providing armoured mobility and fire support for Italian infantry units. (Italian MoD)

Made by Iveco-Fiat, the 23-tonne Dardo was a mechanised infantry fighting vehicle and was introduced to support the M113, which was in Italian service in the beginning of the 1980s but phased out in 1991. (Italian MoD)

The 25mm KBA autocannon serves as the primary weaponry. It can fire 600 rounds per minute on a cycle, and the weapon can hold 200 rounds ready to fire. As a coaxial armament, a 7.62mm MG42/59 machine gun is installed. Finally, one TOW anti-tank missile was installed on each side of the turret of the prototype.

Dardo Infantry Fighting Vehicle Specification	
Model	Dardo
Manufacturer	Iveco Fiat OTO Melara Syndicated Company
Country	Italy
Year	1998–present
Engine	Iveco-Fiat V6 MTCA turbodiesel 512hp (382.2kW)
Fuel	Diesel
Protection	Modular front armour against 25mm APDS
Top Speed	43mph (70km/h)
Range	372 miles (600km)
Crew Capacity	3 (commander, driver, gunner), 6 passengers
Length	22ft (6.7m)
Width	9.8ft (3m)
Height	8.66ft (2.64m)
Main Armament	25mm Oerlikon KBA automatic cannon; 2x TOW ATGM (optional)
Secondary Armament	7.62mm coaxial machine gun; smoke-grenade launchers
Weight	25.7 tons (23.4 tonnes)
Service Branch	Italian Army

The Dardo was fitted with appliqué armour while deployed in Iraq. It carried three crew and a team of six in the passenger area at the rear of the vehicle. (Italian MoD)

Centauro – Wheeled Anti-Tank Vehicle

One of the weapons platforms deployed to Iraq by the Italian contingent in 2003 was the large, eight-wheeled Centauro, dubbed the tank-killer. First entering service in 1991, its main armament was the OTO Melara 105mm gyro-stabilised high-pressure, low-recoil gun equipped with a thermal sleeve and an integrated fume extractor. Of its 40-round capacity, 14 ready rounds were stored in the turret and the other 26 in the hull. The Centauro's secondary weapon was a 7.62mm coaxial machine gun and a second 7.62mm machine gun for anti-aircraft with 4,000 rounds of ammunition.

The Centauro's steel armoured hull is all-welded, and in the baseline configuration is designed to withstand 14.5mm bullets and shell fragments, with protection against 25mm ordnance on the frontal section. The addition of bolt-on appliqué armour increases protection against 30mm rounds. It is also equipped with a chemical, biological, radiological and nuclear (CBRN) warfare protection system, which is integrated with the vehicle's air-conditioning system. The vehicle is also equipped with a four-barrelled smoke grenade launcher mounted on each side of the turret.

The Centauro was the Italians' largest mobile weapon. In one operation in early 2004, a mechanised column of 60 vehicles, including eight Centauro armoured reconnaissance platforms, arrived in the southern zone of Nasiriyah at dawn. The patrol quickly came under attack near the Euphrates River. The Italians responded with the Centauro's 105mm cannons, destroying a building used by Iraqi snipers.

The Centauro was usually employed to escort motor convoys, for wide area control, and for road patrols, but in battle conditions, its commander can find and identify targets without rotating the turret by using the independent panoramic sight. He can also use it to aim the main gun in order for the gunner to engage. This permits the crew to work on engaging more than one target at a time, day and night and in all weather conditions. While the gunner neutralises the first target, the tank commander can find others, identify them one at a time and send the data to the computer. As soon as the gunner has

The Italian contingent deployed the Centauro to Iraq in 2003. This eight-wheeled tank killer was fitted with an Oto Melara 105mm gyro-stabilised, low-recoil gun. The crew carried 40 rounds, with 14 ready rounds in the turret and another 26 rounds in the hull. (Italian MoD)

The Centauro is also equipped with a chemical, biological, radiological and nuclear (CBRN) warfare protection system, which is integrated with the vehicle's air conditioning system. The vehicle has a four-barrelled smoke grenade launcher mounted on each side of the turret. (Italian MoD)

A Centauro 'tank killer' heads an Italian military convoy in central Iraq. The 15-tonne wheeled platform was fitted with a 105mm main gun that can re-engage a target while the commander identifies and selects a second via the Centauro's sophisticated targeting system. (Italian MoD)

eliminated the first target, the targeting computer will turn the gun automatically to the second target, which can be engaged once the loader has finished the loading operation. The Centauro has since been upgraded and retains a leading role within Italian mobile artillery.

Centauro B1	
Model	BMR-600 PP (Porta Personal)
Manufacturer	ENASA
Country	Spain
Year	1979
Engine	Pegaso 9157/8 306hp diesel engine
Fuel	Diesel
Protection	Steel
Top Speed	62mph (100km/h)
Range	373 miles (600km)
Crew Capacity	10 (driver, gunner, 8 troops)
Length	20.17ft (6.15m)
Width	8.2ft (2.5m)
Height	7.7ft (2.36m)
Main Armament	90mm cannon, 7.62mm machine gun, 81mm or 120mm mortar
Weight	16.9 tons (15.4 tonnes)
Service Branch	Italian Army

Spain

In March 2003, the Spanish Prime Minister José María Aznar told Parliament that Spain would not send combat troops to fight alongside the United States and Britain in the invasion of Iraq. However, it did contribute 900 troops for peacekeeping, as well as medical and mine-clearing roles. Their deployment was part of a wider force that included four battalions drawn from the Dominican Republic, El Salvador, Honduras and Nicaragua in a force called the Plus Ultra Brigade. The four Central American battalions were equipped and transported by the US military, and received some specific training in Germany prior to their arrival in the Persian Gulf.

The Spanish expeditionary force based itself in Al-Qādisiyyah. The other central American countries deployed to Najaf in south-central Iraq, near Dīwānīyah. The Coalition objective was that the Brigade would relieve the US Marines at these locations for redeployment to other regions. During their tenure in the region, the Plus Ultra Brigade's troops experienced several hostile clashes with insurgents. A skirmish with insurgents in Najaf in early April 2004 was serious enough to leave one Salvadoran soldier and at least 19 Iraqis dead.

The Plus Ultra Brigade finally dissolved in April 2004, when the newly elected socialist government in Spain and the governments of Nicaragua, the Dominican Republic and Honduras decided to withdraw their troops. The lack of public support for the deployment and the war in Iraq was cited as the main reason, with Nicaragua adding that the cost of the deployment was too much. In December 1998, the remaining Salvadoran unit was reduced to just 200 from its original strength of 390, and a year later was also withdrawn from operations.

The BMR-600 served as Spain's primary wheeled armoured personnel carrier, offering mobility and protection for troop transport. In 2003, the Spanish military deployed a dozen BMR-600s in Iraq. It was capable of carrying out a variety of tasks, including as a command post, as an anti-tank missile carrier, medical evacuation, fire support and troop transport. (Spanish MoD)

Troops entered the BMR-600 via a hydraulic ramp. Its armour was able to protect soldiers from roadside bombs but, as the campaign continued, the IEDs increased in size and lethality. (Spanish MoD)

BMR-600 Armoured Personnel Carrier (Pegaso)

The BMR-600 (Blindado Medio de Ruedas) served as Spain's primary wheeled armoured personnel carrier, offering mobility and protection for troop transport. A domestically produced, medium-weight, six-wheeled armoured vehicle manufactured under licence from France, it has sharp angles to deflect rocket-propelled grenades and can be fitted with additional armour. Twelve were deployed in 2003 by the Spanish military in Iraq.

The BMR-600s was capable of carrying out a variety of tasks, including as a command post, anti-tank missile carrier, medical evacuation vehicle, fire support unit and troop transport. Originally equipped with a Pegaso 9157/8 306hp diesel engine, it features an amphibious capability, an automated gearbox, a torque converter and independent suspension on all six wheels. It can be fitted with both 7.62mm and .50-calibre machine guns. Optional amphibious equipment includes two hydro jets for underwater travel.

Its hull features spaced armour at the front and is constructed entirely of welded aluminium, with heavier protection at the front. This can provide defence against small-arms fire up to 12.7mm calibre. In addition, the armour on the sides and rear of the vehicle protects against shell splinters and small-arms fire and could also protect soldiers against anti-personnel mines and small IEDS. The machine gunner/radio operator and driver make up the two-person crew, with nine infantrymen able to be carried. A bulletproof windscreen protects the driver, who is seated on the left side of the car, and there are smaller bulletproof windscreens on either side of him (also known as periscopes). An armoured flap can be deployed across the front window, with forward visibility possible through a periscope fixed in the roof. Overhead, a single-piece hatch cover that lifts and swings to the right is sited above the driver's seat. With the exhaust pipe on the right side of the hull and the air intake and outflow louvres in the roof, the engine is located to the right of the driver. The front of the hull has an engine access plate for maintenance and, on upgraded models, the mounted machine gun can be fired remotely.

Pegaso BMR Specification	
Model	BMR-600 PP (Porta Personal)
Manufacturer	ENASA
Country	Spain
Year	1979
Engine	Pegaso 9157/8 306hp diesel engine
Fuel	Diesel
Protection	Steel
Top Speed	62mph (100km/h)
Range	373 miles (600km)
Crew Capacity	10 (driver, gunner, 8 troops)
Length	20.17ft (6.15m)
Width	8.2ft (2.5m)
Height	7.7ft (2.36m)
Main Armament	90mm cannon, 7.62mm machine gun, 81mm or 120mm mortar
Weight	16.9 tons (15.4 tonnes)
Service Branch	Spanish Army

Romania

In June 2003, the Romanian Parliament joined Ukraine in agreeing to send peacekeepers to Iraq. An initial figure of 670 soldiers, including infantrymen, infrastructure technicians, medics and military policemen was approved and deployed in the British sector at Basra in southern Iraq. An uplift was later agreed and headed by Romania's 26th Infantry 'Red Scorpions' Battalion, which used the BTR-80, called the TAB by Romanian forces, as its main protected vehicle. From August 2003, Romania deployed more than 5,200 troops to Iraq in support of Operation *Iraqi Freedom*. Romanian forces provided intelligence support to Multinational Division South East (MNDSE) by conducting reconnaissance and surveillance missions and operating unmanned aerial vehicle (UAV) platforms. The Romanians also provided base security, supply-route security and a quick-reaction force in Basra, as well as providing training and monitoring of Iraqi army units. However, the abduction of Romanian journalists in 2004 changed public opinion on the war. Protests took place against the government and in June 2009, Romanian President Traian Băsescu announced that his troops would be withdrawn.

BTR-80 Armoured Personnel Carrier (TAB)

The TAB was a Soviet-designed BTR-80 built under government licence in Romania and introduced in 1986 as a heavily upgraded modern progression of the BTR-60.

In 2003, the Romanian military deployed 14 TABs (BTR-80s) to southern Iraq. As opposed to the twin petrol-engine configuration of the BTR-60, the BTR-80 operated a single V-8 turbocharged water-cooled diesel engine, delivering more power than its predecessor. The BTR-80's armament includes a 14.5mm KPVT heavy machine gun and 7.62mm PKT (PKTM) machine gun. The crew can enter and leave the vehicle through either the upper hatches in the hull roof or by the side doors, which enable entry and egress even when the vehicle is on the move.

High mobility was ensured through the BTR-80's eight-wheel running gear and all-wheel drive, its powerful diesel engine, an independent wishbone torsion-bar suspension and bulletproof tubeless

Romanian forces of the 341st Infantry Battalion 'White Sharks' in southern Iraq with a BTR-80, which was known by the Romanian army as the TAB B33 Zimbru. (UK MoD)

tyres. The vehicle was also capable of crossing water obstacles without preparation. The hull and turret armour were able to protect the crew from small arms fire but was vulnerable to anti-tank missiles. The BTR-80 operated a crew of three, comprising a driver, commander, and gunner, while accommodating up to seven fully equipped soldiers. An expanded family of BTR-80 combat and logistic support vehicles has been developed around the basic chassis: a battalion commander APC, a nuclear and chemical reconnaissance vehicle, a variety of communications, control and command staff vehicles, an armoured medical vehicle, an armoured recovery vehicle and the 120mm Nona-SVK self-propelled artillery gun. In southern Iraq, the Romanians also used the BTR-80 in the ambulance role.

BTR-80 – Armoured Personnel Carrier (Wheeled) (TAB) Specification	
Model	B33 Zimbru
Manufacturer	Arzamas Machine Building Plant
Country	Romania (Soviet licensed)
Year	1984–present
Engine	Diesel KamAz-7403 260hp
Diesel	Diesel
Protection	Armour (10mm at front)
Top Speed	50mph (80km/h)
Range	430 miles (700km)
Crew Capacity	3 plus 7 passengers
Length	25ft (7.65m)
Width	9ft (2.7m)
Height	8ft (2.43m)
Armament	14.5mm KPVT heavy machine gun or 30mm 2A72
Weight	15.4 tons (14 tonnes)
Service Branch	Romanian Army

A Romanian TAB B33 Zimbru at a parade after the conflict in Iraq. The B33 Zimbru was a licensed domestic-built BTR-80, which allowed Romania to configure it to the specifications of its own armed forces. (Romanian MoD)

Ukraine

The Ukrainian government agreed to send troops to Iraq in June 2003, marking the largest military deployment so far mounted by this young country. More than 6,000 Ukrainians saw service in Iraq during the war, including a permanent presence of 1,600. The force included a specialist NBC battalion as well as a large infantry force supported by BTR-80 protected vehicles. Throughout its deployment, the Ukrainian contribution was restricted to a peacekeeping role, although there were instances of combat with Iraqi insurgents. On 9 December 2008, Ukraine formally withdrew its last forces from Iraq, ending its participation in the conflict. Ukraine's involvement in Iraq was strongly opposed by the Ukrainian population, and was seen both within and outside the country primarily as an effort to deflect attention away from President Leonid Kuchma, who was involved in a political scandal. Public opposition to war increased in 2004 following the Ukrainian troops' hasty retreat and loss of their base at Kut to insurgents. The incident infuriated Coalition leaders and led to a reassessment of Ukrainian activities in Iraq. Following the Ukrainian elections of 2004, Kuchma's successor, Viktor Yushchenko, announced the withdrawal of Ukraine's forces. During the conflict as a whole, 18 Ukrainian soldiers were killed.

BTR-94

Ukraine, like Romania, also produced BTR-80s under licence and exported them. Its mechanical profile was the same as the Romanian examples. The country's own modification after independence from the USSR produced the BTR-94, with a larger turret and improved main armament. As well as a 7.62mm PKT (PKTM) machine gun, there was the BAU-232 remote-operated weapon station, which was created

Ukrainian Army eight-wheeled BTR-94 armoured personnel carriers travel along a highway from Al Kut to As Suwayrah during operations in Iraq. (US DoD)

Ukrainian Army BTR-84 armoured personnel carriers arriving at the Iran/Iraq border crossing located at Badrah, Iraq, during Operation *Iraqi Freedom* as Ukrainian forces take command of the local area. (US DoD)

US Marines aboard a Humvee pass Ukrainian Army soldiers aboard a BTR-94 (8x8) APC, on the highway from Al Kut to As Suwayrah in Iraq. (US DoD)

by the SOE Kharkiv Morozov Machine Building Design Bureau. The gunner is situated underneath the turret ring, while the crew can enter and exit the vehicle through either the upper hatches in the hull roof or side doors, even when the vehicle is on the move.

In southern Iraq, the Ukrainian BTR-94s, mounting the 14.5mm KPVT heavy machine gun, were distinctive in their grey and sand camouflage. The Ukrainian contingent was drawn into its first major battle in early April 2004, when it faced off in the city of Kut against the Mahdi Army, commanded by the Shia cleric Muqtada al-Sadr. In the resulting clashes, Ukrainian soldiers lost one dead and six wounded in an anti-tank grenade attack, despite reportedly killing more than 150 insurgents. In the face of this heavy resistance from the Mahdi Army, Ukrainian forces withdrew from the city on 7 April. The retreat, against the orders of the US military commander in Baghdad caused tension within the Coalition, even though al-Sadr's militia had proved to be a far stronger military force than hitherto believed. By 8 April, the Mahdi Army was effectively in control of Kut and Kufa.

BTR-94 - Armoured Personnel Carrier (Wheeled) Specification	
Model	BTR-94 armoured personnel carrier (wheeled)
Manufacturer	Malyshev Factory
Country	Ukraine
Year	1994–present
Engine	Diesel KamAz-7403 260hp
Diesel	Diesel
Protection	Armour (10mm at front)
Top Speed	53mph (85km/h)
Range	370 miles (595km)
Crew Capacity	3 plus 7 passengers
Length	25.09ft (7.65m)
Width	9.51ft (2.9m)
Height	9.18ft (2.80m)
Armament	14.5mm KPVT heavy machine gun or twin 23x152mm 2A7M cannon
Weight	14.9 tons (13.6 Tonnes)
Service Branch	Ukrainian Army

Chapter 5

The Legacy of War in Iraq

The military legacy of war in Iraq was a new generation of armoured protected vehicles and a fresh approach to protected mobility in counter-insurgency operations, developed to meet the growing threat of insurgents' roadside bomb attacks. For the US and the other countries which sent troops to serve in Iraq, the overall lesson learned was that 'force protection' was a major consideration when deploying troops amid a counter-insurgency operation. Tanks and armoured personnel carriers had been regarded as presenting an 'aggressive posture' that escalated the temperature in affected areas. A similar situation was when the British used armour in Northern Ireland after light-skinned vehicles failed to protect soldiers from roadside attacks. While tanks were not deployed, the British did upgrade Land Rovers with heavy armoured packages. As the tempo of attacks eased, these were in turn replaced with lighter vehicles, known as Snatch Land Rovers. In Basra, however, these vehicles had proven to be totally inappropriate.

In Iraq, armoured vehicles provided critical capabilities during the initial drive on Baghdad and Basra. They also quickly proved vital to counter-insurgency, the unintended next phase of operations. One of the most successful new systems deployed to Iraq was the Stryker armoured vehicle, which was wheeled rather than tracked for increased mobility and speed. Although more lightly armoured than

US Army soldiers assigned to the 172nd Stryker Brigade Combat Team, A Company, 2nd Battalion, 1st Infantry Regiment, conduct a reassurance patrol at Mosul in northern Iraq. The Kurdish community in the north supported the US-led invasion. (US DoD)

A private security company helicopter flies above the Green Zone in Baghdad, where Coalition troops and diplomats were stationed, minutes after a car bomb exploded. (US DoD)

an Abrams tank, the Stryker has survived hundreds of hits by rocket-propelled grenades while offering soldiers greater flexibility when on patrol in dangerous environments.

The inventory of Army equipment in 2003 was conceived for fighting conventional adversaries in circumstances where secure and contested areas are well defined. However, counter-insurgency campaigns rarely develop with any certainty, and Iraq was no exception. In fact, a central feature of the Iraq conflict was the inability of defenders to fully secure cities and countryside against an elusive enemy. Coalition forces remained in continuous danger whenever they left guarded compounds and, as such, 'protected vehicles' became a necessity. Iraq demonstrated to extremists around the world how effective terrorist tactics can be against formed units.

US and Coalition forces discovered very quickly that their vehicles enjoyed limited protection against counter-insurgency operations. In Baghdad, the Humvee needed significant armoured upgrades after insurgents attacked vehicles with IEDs. Such was the size of these roadside bombs that many of the armoured vehicles that took part in the assault were assigned extra armour packages. When it arrived in October 2003, the Stryker was also fitted with extra protection.

US military commanders had prepared for war, not a volatile peace-support operation, and as such, had been left exposed. While the US and Multi-National Forces sustained their campaign with a new generation of Mine Resistant Ambush Protected (MRAP) vehicles, the legacy for the wider Middle East was chaos. President Bush's decision to invade Iraq resulted in regional instability and left thousands of displaced persons, many of whom would go on to seek refuge in the West. Bush had claimed that Saddam had weapons of mass destruction that he could use against the West, a statement that was widely challenged before the war. Just months after the mission was launched, it was revealed that the WMDs, did not exist and as such, Saddam and his military commanders were not the threat that the White House had claimed. In the meantime, Baghdad had been destroyed, infrastructure broken and hundreds of civilians killed, leaving families homeless and men of fighting age frustrated. After the initial invasion, more than 35 countries poured troops into Iraq to support a peacekeeping operation, but the mission quickly became kinetic and a demand for 'protected mobility' spiked. The insurgents continued to attack Coalition vehicles, forcing the US and its allies to respond with bigger and better armoured vehicles and tanks.

Ironically, in Iraq's war with Iran in 1980, the United States had been an ally of Saddam. At the time, Washington knew quite well that Saddam was a brutal mass murderer, but to the White House he was preferable to an Iranian Islamist government. The US military gave Saddam Hussein's commanders targeting information to strike at Tehran and military chiefs poured weapons into Baghdad. President Reagan's administration removed Iraq from the list of state sponsors of terrorism so that US weapons could flow unimpeded into Iraq.

Two decades later, relations between the US and Iraq had entirely reversed. Even after Operation *Desert Storm* evicted Saddam's forces from Kuwait in 1991, diplomatic relationships between the two countries were not re-established. President George W. Bush went on to claim that Iraq was one of three countries, the others being Iran and North Korea, forming an 'axis of evil'. While there was some degree of opposition in the United States, the American public generally supported the President's plan

Two UH-60 Black Hawk helicopters land next to a US Army Humvee inside the International Zone in Baghdad, known as the Green Zone. The crossed sabres, made from Iranian armour and helmets from Iranian soldiers, are part of the parade ground that was used by Saddam Hussein when reviewing his army. (US DoD)

to invade Iraq in 2003. In the build-up to the war, public opinion showed that many people accepted claims that Iraq had supported al-Qaeda in the terrorist attacks on the United States of 11 September 2001. Others feared another attack on mainland America and feared Saddam would arm al-Qaeda with weapons of mass destruction. In Europe, thousands marched against the war, but Bush had made his decision and British Prime Minister Tony Blair agreed to send armour and thousands of troops to join the invasion. Australia and Poland also stepped forward with ground troops in support.

The first shots in the 2003 war started on 19 March with an overwhelming show of military might, described by the unforgettable phrase 'shock and awe' in which Coalition bombs levelled parts of Baghdad. America's key aim in invading Iraq was to depose Saddam Hussein, remove his alleged WMDs and liberate the population from the tyranny of a dictator. In northern Iraq, US forces supported Kurdish fighters in Operation *Viking Hammer*. The aim here was to eliminate the Kurdish Salafist group Ansar al-Islam and dismantle the Islamic Emirate of Byara that this separatist movement had set up. Established in 2001 by former al-Qaeda members, with cross-border support from Iran, Ansar al-Islam imposed a strict application of Sharia in villages it controlled. Kurdish Peshmerga forces as well as armed civilians in Kirkuk, Mosul, and Diyala all co-operated with US forces and helped to liberate the northern areas and eastern areas of Iraq.

While the military offensive was comprehensive and executed with clinical professionalism, Washington had failed to draw up a post-war plan for reconstruction and development. Added to the fact

Coalition forces respond to a car bombing in south Baghdad as a second car bomb is detonated, targeting those responding to the initial incident. The attack, aimed at the Iraqi police force, resulted in 18 casualties. Just months after the invasion, these attacks soared in frequency. (US DoD)

that the Coalition needed ever more troops just for pacification purposes, there was little understanding of the myriad cultures in Iraq and no plan to help the population as a whole.

The primary objective of Operation *Iraqi Freedom* had been to oust Saddam's regime. However, the US-led Coalition in Baghdad quickly became a 'colonial power' as Washington formed what was to be called the Coalition Provisional Authority (CPA). It was headed by former US diplomat Paul Bremer, who immediately decided to disband the Iraqi Army, a force of 300,000. The impact of this controversial decision was immense; unemployed soldiers lost their income and could not support their families. Washington planned to build a New Iraqi Army, but a significant delay in delivering a small weekly financial retainer to former soldiers simply ensured that many joined the insurgency springing up. As hostile numbers soared, Iran moved to supply the insurgents with weapons and explosives in exchange for intelligence on the US. The situation on the ground deteriorated very quickly as looting and violence erupted across the country. The growing insurgency, plus the breaking out of an effective civil war between Sunni and Shia militias destroyed what temporary order that had been established.

In Saddam Hussein's Iraq, membership in the Ba'ath party was the standard requirement for state employment, creating a hegemonic party with a foothold in every public institution and all corners of society. The party had immense powers, which it used to engage with the state security apparatus in widespread unauthorised killings, enforced disappearances, torture and horrific human-rights violations during three decades of brutal rule. With this in mind, the CPA strove to implement a de-Ba'athification process intended to rid the country of the Ba'ath party's influence. Thousands of individuals were dismissed based on their rank within the Ba'ath party hierarchy, rather than on their conduct, and resulted in the total collapse of the Iraqi civil service.

A 'hearts and minds' campaign was mounted by US forces alongside an ongoing operation to hunt down the Iraqi commanders and senior politicians who had fled as the Coalition arrived in the city. Saddam was also missing. In their task to find top commanders, dubbed 'the most wanted', American soldiers searched neighbourhoods and stopped cars and lorries, often finding an AK-47; most Iraqi homes had at least one gun and these were particularly widespread. However, in the tense atmosphere of house searches, the understandably nervous troops often burst in and verbally abused or manhandled the head of the family, the father. In Iraqi culture, the man is the head of the family and his place is fundamental to the wider respect of the family within the community. This was a difficult balance, as soldiers needed to find and detain those on the 'most wanted' list but also needed to maintain support from the community.

A British soldier attached to a US airborne unit suggested that instead of smashing down doors, the US soldiers should consider asking the local imam to join them and explain to the community what they were doing, while thanking the head of the household for his co-operation. There was resistance, as this meant the element of surprise would be lost, but in the longer term it would build a relationship in which the local population may be persuaded to trust the soldiers. The tactic of using an imam was not easy to establish but when it was achieved, it showed signs of success. Even so, it was abandoned after two weeks as commanders demanded more surprise raids and more 'shock and awe'. On the streets, dozens of homeless and disabled children were now rounded up by American forces and put into a home that offered bed, food and clean clothes. Iraqi families were furious, however; they had put these disabled children onto the street because they could not earn a wage and in their culture, rightly or wrongly, they would survive as vagrants or pass away. The lack of cultural awareness impacted heavily on the ability to build a hearts and minds campaign, instead broadening the gap of distrust and anger between the Iraqi public and the invading force. The poor management of house searches resulted in many disaffected Iraqis joining the insurgency; in its insensitivity, the Coalition had inadvertently become the 'recruiting sergeant' for the terrorists. Attacks thus increased as insurgents dug bombs into the road. Although

Above: Brass shell casings lie scattered in the road as US Army soldiers from Alpha Troop, 1st Battalion, 75th Cavalry Squadron, 101st Airborne Division, provide security after an attack on Coalition forces in Baghdad, Iraq. (US DoD)

Left: A US Army soldier of Alpha Troop, 1st Squadron, 75th Cavalry, 2nd Brigade Combat Team, 101st Airborne Division, provides security while on patrol in the Gazaliyah district of Baghdad in September 2008. (US DoD)

enhanced armour was deployed to address this and a new generation of protected vehicles made its appearance across Iraq, the insurgents responded by making bigger bombs.

Late in 2003, Saddam Hussein was captured during Operation *Red Dawn*. Three years later, he was executed after a trial in Baghdad, but violence endured nonetheless, taking on ever more forms.

The professional approach of the US military was generally beyond question, but when it came to maintaining the peace, the understanding of local cultures and what might be called 'emotional intelligence' was sometimes lost on senior commanders. By the end of 2003, Shia cleric and militia leader Muqtada al-Sadr was known by the Coalition to be a key player in attacks against Western forces and was on the most wanted list. He and a number of his cohorts had been located in the city of Karbala by US intelligence. In the middle of the night on 14 October, a contingent of US soldiers flew from Baghdad to Karbala, headed by General Ricardo Sanchez, the head of Coalition forces. It was revealed that Sadr and his men were hiding in the golden mosque and, after an hour of planning, the general made his decision to destroy the structure with a JDAM (Joint Direct Attack Munition) a huge bomb that can be placed with pinpoint accuracy. The general asked if there were any questions, to which one officer replied by suggesting that the 'collateral fallout' of the JDAM would hit a nearby school, cause immense cultural as well as physical damage and risk killing innocent civilians. He asked the officer, a British soldier, what he would do. 'We would gas the mosque and send in the Iraqi Police, supported by Coalition forces. The decision was made not to bomb the mosque.

A US Marine Corps M1A1 Abrams Main Battle Tank, 2nd Tank Battalion (BN), fires its main gun into a building to provide suppressive counter fire against terrorists who fired on other Marines during a firefight at Fallujah in Al Anbar Province, Iraq. (US DoD)

A cloud of smoke and dust envelops a soldier seconds after he fired an AT-4 rocket launcher at an insurgent position during a firefight in Baghdad's Adhamiyah neighbourhood, which ended with one insurgent dead and three captured. (US DoD)

In Fallujah, an hour west of Baghdad, hundreds were killed when insurgents seized the city in April 2004. The US Marines sent armoured Humvees and Abrams to regain control in what became a brutal battle. As the conflict continued, the nation-building that the US planned collapsed, and instead, the Coalition and Multi-National Force focused on training the New Iraqi Army as its main route towards an exit strategy. In 2007, the security situation started to improve when the US deployed an additional 30,000 troops to Iraq in the famous 'surge'. Four years later, however, Bush's successor, Barack Obama, decided to withdraw the majority of US troops in the face of worsening political relations with the Iraqi government.

By 2009, most nations had withdrawn their forces from Iraq and the pressure on the US to maintain a presence on its own increased. The US plan to establish a democratic government in Iraq, a country that had not known elections under Saddam, and the brutal nature of the counter-insurgency operation gave oxygen to a new insurgency called Islamic State (IS). This organisation had emerged from the remnants of al-Qaeda in late 2011 and changed its name to the Islamic State of Iraq and Syria (ISIS). ISIS launched an offensive on Mosul and Tikrit in June 2014 and on the 29th of that month, ISIS leader Abu Bakr al-Baghdadi announced the formation of a caliphate stretching from Aleppo in Syria to Diyala in Iraq.

After almost a decade of fighting to deliver democracy in Iraq, America was forced to fund further new equipment and training to allow the New Iraqi Army to fight IS in an operation in which the US directed and assisted the Iraqis. By 2012, the Iraqis were equipped with the latest US military equipment

Suicide bombings in Iraq since 2003 have killed thousands of people, mostly Iraqi civilians, and were launched as part of a growing insurgency. This is a familiar tactic in other armed struggles, but their frequency and lethality in Iraq was unprecedented. (US DoD)

including Abrams tanks, armoured protected vehicles such as the Caiman and armoured Humvees. Thus trained, the Iraqis used their Abrams, Cougars and many other armoured vehicles in their battle to retake the city of Mosul and drive out IS into Syria.

Syria had problems of its own; after popular revolts in Tunisia, Egypt and Libya, an attempt was made to oust long-time Syrian dictator Bashar al-Assad, who simultaneously found his own forces besieged by IS and another Syrian rebel army supplied by Turkey. Although US support helped the Iraqi Army and the Syrian opposition grind down IS, it was Russian intervention on behalf of al-Assad that not only tipped the balance of local relations in this part of the Middle East, but emboldened Russian President Vladimir Putin to embark on a series of military adventures of his own, which would characterise the following decade.

Though US President Donald Trump (2017–21) loudly claimed credit for defeating ISIS, his successor Joe Biden announced that he would end the US combat mission in Iraq by the end of 2021, leaving a small number of American troops (an estimated 2,000) in an advisory and assistance role.

All through these conflicts, Iran has been the long-standing nemesis of Washington. The US's support for Iraq in the 1980s in the first place was part of the fundamental aim of countering Iranian influence in the Middle East, but the subsequent defeat of Saddam brought Iran into a stronger position locally, aided by ethnic as well as religious links. In the war of 2003 and thereafter, the Coalition came face to face with bombs that Tehran had passed to the Iraqi insurgency.

A New Iraqi Army (NIA) convoy rolls through Mosul in northern Iraq on a routine patrol with upgraded armoured Humvees and Mine Resistant Ambush Patrol vehicles (MRAPs). (US DoD)

More than a dozen Iraqi political parties have links to Iran, which funds and trains paramilitary groups aligned with these parties. Some paramilitary groups under the umbrella of Iraq's Popular Mobilisation Forces have pledged allegiance to Iran's supreme leader, Ayatollah Ali Khamenei. These groups have used violence to crush opposition to Iranian influence and constantly campaign to expel the remaining US forces based in Iraq. Since the Islamic State was driven out in 2017, Iran-backed militias have frequently launched attacks on US troops who remain in Iraq in a non-combat, advisory capacity. Some of the most advanced protected vehicles, including Abrams tanks, were seized from Iraqi forces by militia groups and passed to Iran.

Today's Iraqi government thus finds itself heavily influenced by Iran, to the concern of the US State Department, which has made public its concern about the undermining of the stability and the integrity of Iraq's national democratic institutions. The wider implications of Iranian geopolitics are also never far from the surface, specifically the country's open links to Hamas and Hezbollah. Not long after war broke out in Gaza in October 2023, Iraqi airspace saw regular flights of hundreds of projectiles from Iran towards Israel. Thus is Iraq caught in the crossfire of what by 2024 is effectively the newest regional war. More than 20 years after the US invaded Iraq in 2003 to remove Saddam Hussein, roughly 2,500 US troops are still based in the country, primarily to counter the not insignificant remains of Islamic State.

Then and now, the armoured vehicles employed to transport the American and Coalition troops into the heart of Iraq and then to defend them against attacks more deadly than any encountered during the invasion itself, have played a central role; as outlined in detail in this comprehensive account, their armament, equipment, performance and protection have been constantly upgraded and improved throughout.

Right: Iraqi forces parade their new Abrams tanks during the Iraqi Army Day celebration in the International Zone in Baghdad. The Abrams and other armoured vehicles significantly increase Baghdad's capability. (US DoD)

Below: A US tactical M-ATV during a co-ordinated patrol with Turkish military forces along the demarcation line outside Manbij, Syria. (US DoD)

Burned-out cars rest on the outskirts of Mosul during the fighting against Islamic State (ISIS). The major battle here was initiated by Iraqi government forces with allied militias drawn from the Kurdistan Regional Government to retake the city from ISIS, which had seized it in June 2014. (US DoD)

A US mobile construction battalion takes a break to hand out cookies to local Iraqi children as part of their 'hearts and minds' operation while helping to build a school and improve the water, electrical and sanitation facilities at a small Bedouin village on the outskirts of Najaf, Najaf province, during Operation *Iraqi Freedom*. (US DoD)

A USMC Humvee pulls up to form a checkpoint on Route Ethan in Fallujah as a US Marine convoy prepares to pass through the area. (US DoD)

A New Iraqi Army parade of armour-upgraded Humvees supplied by the US as part of the majority overhaul of Iraqi military assets. (US DOD)

Other books you might like:

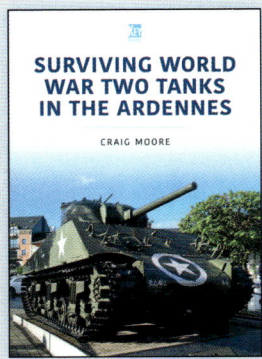

Military Vehicles and Artillery Series, Vol. 4

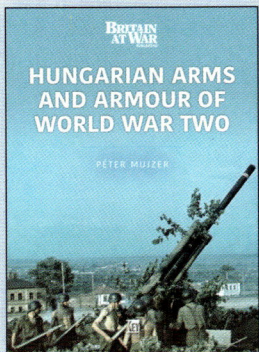

Military Vehicles and Artillery Series, Vol. 5

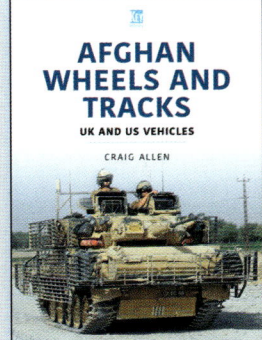

Military Vehicles and Artillery Series, Vol. 6

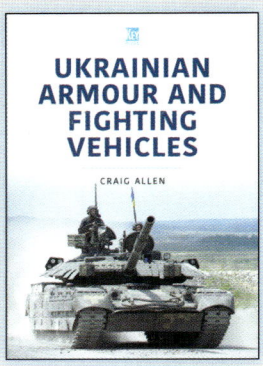

Military Vehicles and Artillery Series, Vol. 7

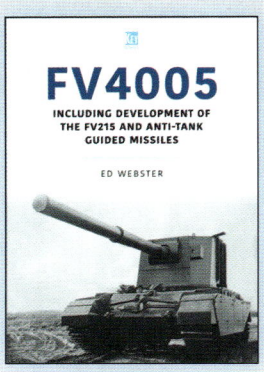

Military Vehicles and Artillery Series, Vol. 8

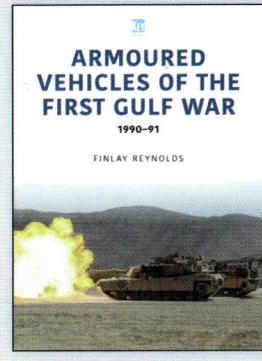

Military Vehicles and Artillery Series, Vol. 9

For our full range of titles please visit:
shop.keypublishing.com/books

VIP Book Club

Sign up today and receive
TWO FREE E-BOOKS

Be the first to find out about our forthcoming book releases and receive exclusive offers.

Register now at **keypublishing.com/vip-book-club**

Our VIP Book Club is a 100% spam-free zone, and we will never share your email with anyone else. You can read our full privacy policy at: privacy.keypublishing.com